W9-COD-092

THE STARS IN OUR POCKETS

THE STARS IN OUR POCKETS

GETTING LOST AND SOMETIMES FOUND IN THE DIGITAL AGE

HOWARD AXELROD

BEACON PRESS
BOSTON

BEACON PRESS
Boston, Massachusetts
www.beacon .org

Beacon Press books
are published under the auspices of
the Unitarian Universalist Association of Congregations.

23 22 21 20   8 7 6 5 4 3 2 1

This book is printed on acid-free paper that meets the uncoated paper
ANSI/NISO specifications for permanence as revised in 1992.

Text design and composition by Kim Arney

Library of Congress Cataloging-in-Publication Data
Names: Axelrod, Howard, author.
Title: The stars in our pockets : getting lost and sometimes
found in the digital age / Howard Axelrod.
Description: Boston : Beacon Press, [2020] | Includes bibliographical references.
Identifiers: LCCN 2019020804 (print) | LCCN 2019980588 (ebook) |
ISBN 9780807036754 (hardcover) | ISBN 9780807036761 (ebook)
Subjects: LCSH: Axelrod, Howard | Technological innovations—Social aspects. |
Information technology—Social aspects. | Civilization, Modern—21st century.
Classification: LCC HM846 .A94 2020 (print) | LCC HM846 (ebook) |
DDC 303.48/33—dc23
LC record available at https://lccn.loc.gov/2019020804
LC ebook record available at https://lccn.loc.gov/2019980588

In Memory of Oliver Sacks

If he forced himself to contemplate the constellations night after night and year after year, following their progress, their returns along the curved tracks of the dark vault, he, too, would perhaps gain in the end the notion of a continuing and unchangeable time, separated from the labile and fragmentary time of terrestrial events. But would attention to the celestial revolutions be enough to stamp this imprint on him? Or would not a special inner revolution be necessary, something he could suppose only theoretically, unable to imagine the palpable effects on his emotions and on the rhythms of his mind?

—ITALO CALVINO, from *Mr. Palomar*

CONTENTS

THE STARS IN OUR POCKETS

INTRODUCTION

Inner Climate Change

A FEW YEARS AFTER COLLEGE GRADUATION, I moved into a house at the dead end of an unmaintained dirt road, deep in the woods of Vermont's northeast kingdom. I had no TV, no cell phone, and no computer. The newspapers piled by the woodstove told of lost dogs, bingo nights, and spaghetti dinners already years gone, and by that first December I needed those "cutest darn mutts" less as ghostly company than as kindling, their tracks through the snow into newsprint finding purpose again as crumpled balls, then as glowing sea anemones flaring over the previous night's embers. Every two weeks or so, if the road wasn't snowed in, I'd make the drive down into the village of Barton to stock up on provisions. There I'd catch the dazzle of magazine covers by the produce aisle, the celebrity gossip like a strange mute party that needed your attention to turn the sound back on. But usually I went days without hearing, speaking, or seeing a word, other than the wisdom on my teabag—Variety Is the Spice of Life!—and the ingredient list on the back of my soup.

If your Luddite alarm is flashing an icon of a misanthrope brandishing a walking stick at technology's evils, I don't blame you. I did grow an absurd beard, and my cousin Mitch announced me at Thanksgiving as "the Unabomber." But I wasn't a misanthrope, and my retreat to Vermont, which started in 1999, had nothing to do with technology. It had to do with a question I'd been asking myself since being blinded in my right eye five years earlier—a question that felt both elusive and ubiquitous, so much a part of my life I couldn't avoid it but also couldn't grasp it, a question no ophthalmologist could offer any clues to. *What shapes the way I see?*

Even that wasn't the right question, not exactly. It had more to do with how the air felt around me as I walked up to class, or stepped down the two wide stairs into the dining hall, with how friends felt closer in space since the accident but conversations felt farther away, with how time felt slower and more fluid, and with how there was some part of me, a deep abiding part I didn't know very well, that felt exactly the same. Really, the question was about how the world entered me and how I entered the world, but I couldn't find the shape of the question, let alone its answer, while being surrounded by it.

The accident happened on a beautiful May afternoon, two weeks before the end of my junior year at Harvard. During a game of pickup basketball, in a scuffle for a loose ball, a boy's finger darted into my eye past the knuckle. Time went slow motion, space seemed to collapse, and his fingernail sheared my optic nerve, the cable that connects the back of the eyeball to the brain. Blinded in that eye, I lost peripheral vision, depth perception, and also a kind of firmness in the world, an unquestioning confidence in my sense of reality. After ten days convalescing at home, I returned to school for final exams, but everything looked and felt different. I stumbled often on the stairs, knocked over glasses of grape juice in the dining hall, would turn to my right with a full tray and crash into someone who had materialized out of thin air. Perhaps strangest of all, walls no longer looked solid. I noticed it the afternoon I was studying for my Shakespeare exam. A chair scraped on the floor above me, and when I looked up, I couldn't tell where the brown molding ended and where my ceiling began. There was a kind of hovering, a margin of error—the ceiling unwilling to commit to a precise location. The wall behind my desk didn't look particularly solid either.

Everything in the room had a subtle floating quality, as though I'd passed through a fairy-tale portal into a dress rehearsal for reality, as though nothing was really happening in the material world. Instead of going to breakfast, I began to watch the morning from my bedroom window, the patterns of backpacks and jackets shifting like schools of fish. I began to wonder about blind spots, about the tides guiding my friends' daily motions, about the tacit assumptions we took to be natural laws. And I began to be afraid that every morning the world would look unfinished to me, every line the ghost of a line, as though what I was seeing had already happened somewhere else.

So, not long after graduation, I moved to the woods of northern Vermont. My hope was that whatever shaped the way I saw would knit together again like so many broken bones. The house had no clocks, and I began to wake with the sun, to know the progress of the seasons by its angle in the sky. The woods around the house expanded; the nearest town drifted far away. On my daily snowshoe treks through the trees, I began to see and hear things I'd never seen or heard. The fibers of a maple tree constricting and keening in the cold. A camouflaged black-capped chickadee, its feathers tufting in the wind on a snowy branch. The days slowed. My attention opened, a windless pond taking in the trees on its banks. My memory opened too. I could snowshoe my way back to the house through drifted snow and miles of unmarked trees, ignoring my tracks, simply remembering the crumbling stone walls and the landmarks of birches and spruce. As I unloaded groceries from the village market, the songs that had been playing on the overhead speakers would follow, offering a kind of souvenir map of my market experience. When I lifted a package of ramen from the shopping bag one afternoon, Elton John's "Rocket Man" unspooled after it like a thread through the aisles: "I miss the earth so much I miss my wife," I'd been reaching for the ramen at the end of aisle 3; "It's lonely out in space," I'd rounded the potato chip display; "And I think it's going to be a long, long time, till touchdown brings me round again to find . . . ," I'd been reaching into the cooler for the orange juice, where I'd paused to listen, "I'm not the man they think I am at home. Oh, no, no, no . . ."

Maybe this happened because of my long daily walks in the woods and my brain's new habit of absorbing everything it could as reference points. Maybe it happened because I didn't hear music anywhere outside the market, and "Rocket Man" is a weirdly memorable song. Either way, everything I encountered—or didn't encounter—was quietly altering my sense of time, my sense of place, and the quality of my attention and memory. What I was experiencing was changing *how* I was experiencing. Perhaps even stranger, there was no way to feel those changes being made; I could only register they *had* been made once my ramen became a kind of radio.

There in the kitchen, among foods that were songs and songs that were really a map, it began to dawn on me that the truth about perspective is embedded in the word's etymology. It comes from the Latin, *perspicere*, which doesn't mean *to look from* but *to look through*. Perspective isn't a place outside you, a vantage point you can step away from and inspect from afar. It's a lens you can't see because it's a part of *how* you see. It's your inner eye, the complex lens through which you look through your physical eyes and see reality. And it keeps changing as a function of how you're living, adapting to your daily habits and needs, using everything it can to navigate the deep snows and miles of unmarked trees, even when you're just shopping in an eight-aisle village market, tapping your thumbs on the shopping cart to Elton John.

IN HIS MEMOIR *On the Move*, the neurologist Oliver Sacks writes that the only modern theory to have excited him as much as Darwin's theory of evolution is Gerald Edelman's theory of neural Darwinism. Experts used to assume that the brain was fairly rigid, fully programmed at birth, but the Cliffs Notes to contemporary understanding started in 1949 with a phrase known as Hebb's rule: "Cells that fire together wire together." In other words, your particular experiences, and especially your habits, shape the connectivity of your brain by determining which neurons repeatedly cause other neurons to light up. Those lighting-up patterns form particular constellations of neurons among your neuronal population. This is where Edelman comes in. As Sacks explains in *On the Move*:

> Edelman conceived, almost from the start of his career, that processes analogous to natural selection might be crucial for individual organisms . . . with life experiences serving to strengthen certain neuronal connections or constellations in the nervous system and to weaken or extinguish others. . . .
>
> He saw his own work as the completion of Darwin's task, adding selection at a cellular level within the life span of a single individual to that of natural selection over many generations.

In other words, natural selection happens on both sides of your eyes. Based on your habits and experiences, certain populations of neurons get selected and their connections grow stronger, while others effectively go the way of the dodo bird, their synaptic links weakened to the point of uselessness. For example, studies show that the visual area of the brain begins to process auditory and tactile information within five days of the loss of vision, even if that loss is due to a blindfold. Parts of the visual cortex fall dormant while others start to use the way sound travels through space and the touch of wind patterns on your skin to help orient you in your surroundings. The original neuronal connections stop firing because they're not being used, and they adapt by rewiring—and the more your brain uses the new connections, the stronger they become. Of course, if you take off the blindfold, the dodo bird gets resurrected: vision works in the usual way and the original neuronal connections start firing again. But if you can't take the blindfold off, the wiring of your brain adapts.

This is what happened to me after my eye accident. Back at college, I found myself hearing not just sounds but space—the shifting position of pedestrians behind me in the scuffing of footsteps on the narrow brick sidewalk, the distance of an approaching car in the wind-rush coming down Mass. Ave. It was as though my senses had quietly become physics majors, seamlessly calculating relationships between sound, velocity, and distance, and the answers unfolded in my brain as a kind of three-dimensional map, one my eye was consulting without my knowing it. Sometimes it spooked me to know a car would dart out of an alley before I knew how I knew. The neuronal populations in my brain were adapting to the new climate projected by blindness in one eye—no depth perception, diminished peripheral vision—as though a real change in the physical climate had occurred, just as entire populations of species have always adapted to changes in their climates. In the forests of Finland, for instance, the tawny owl has become more brown-plumed than white-plumed over the last several generations as temperatures have warmed, snow has melted, and the owl has needed to camouflage itself among the forest's bare branches. In the same way that natural selection helped make the tawny owl less visible by selecting for traits that are adapted to its new climate, neural Darwinism was helping me "to see" by selecting for neuronal connections adapted to mine.

Oliver Sacks loved Edelman's theory because it helped explain the wonder of what he'd encountered in his patients, their neurological impairments and amazing neurological adaptations, and because it gave him "the feeling of having been liberated from decades of epistemological despair—from a world of shallow, irrelevant computer analogies into one full of rich biological meaning, one which corresponded with the reality of brain and mind." As an enormous Darwin fan, Sacks also likely loved the continuity between the world outside us and the brain inside us—that the adaptive mechanism behind the hummingbird's long beak and the giraffe's long neck was also working inside the human brain to help each of us, whatever our unique perceptual climate, to reach the necessary sensory and cognitive nourishment to orient ourselves in the larger world. I loved the theory for my own reasons: it helped explain my sensory disorientation and reorientation after my accident; it helped explain my experiences in the woods and how Elton John had snuck his way into my ramen; and it helped explain what happened in the years after my return from the woods, when I moved back to Boston and right into the heart of the Digital Age.

WHEN I RETURNED, the sidewalk had changed. It was 2001 and people now walked talking into their hands. Cell-phone use had doubled in the two years I was away. There were terrorist threats, rumors of WMDs, and a year and a half later the Iraq War began. Like everyone else, I wanted to stay informed, to be connected and reassured. But the 24-hour news cycle felt wildly loud to me, its own strange climate of anxiety rather than clarity, clamor rather than conversation, the whole country in need of a teacher, it seemed, to turn off the lights and quiet everyone down. Soon the golden era of social media began, with Friendster in 2002, LinkedIn in 2003, Facebook in 2004, and Twitter in 2006 (by which time Blackberrys were already called Crackberrys), and in 2007, Apple introduced a new product: its touchscreen smartphone.

I felt like Rip Van Winkle—not gone for two years but asleep for twenty. Like I'd woken to a new kind of world entirely, a magical world where people could glance into their hands and see people far away, and yet, because of an optical quirk in the atmosphere, where they had a hard time making out the people in front of them, and so were uncertain whether they were never alone, or always alone, or somehow both. A world where urgent flashes of danger and outrage sparked through the mist at all hours, and where your sense of where you were and who you were required paying attention to these flashes, and flashing your own in return, so you wouldn't lose yourself in the strange way the air played tricks on you in this magical world, the nearby gone foggy and faraway, the faraway suddenly sharp and nearby, where you needed to tap out signals to remind yourself that you existed, to remind yourself that others could see you, to remind yourself that you were not lost. Above the Boston skyline so familiar from my childhood, a few stars still adorned the sky at night, decorative and nostalgic, a reminder of ancient mariners on the seas, but the real stars now were in your hands—that's what you followed to get you through the storm.

Maybe that's what was strangest—this magical realm was superimposed on my hometown. The sidewalk, the bus I took to work, the Starbucks around the corner—each seemed its own version of a ghost town, inhabited by people who were there but who also weren't. Every morning and evening on the bus, I boarded to rows of wires dangling from ears, screens cradled in hands, as though people needed life-support systems to survive Earth's new atmosphere. Oddly, the air did feel thinner than it had in the Vermont woods, as though we were all no longer connected edgewise, as though we were no longer in the same place. If a young man walked straight towards me on the sidewalk, lost in his hand, I'd keep walking towards him. Why should I be responsible for his attention? The second after he veered off, guided by an impressive internal alert, I'd feel like a fool. I was a sidewalk vigilante of time/space morals, of old woods etiquette? How had I appointed myself to such a post—and what did it even mean?

Mostly what I felt was an unfamiliar loneliness. Not just because the city had changed in my years away, which was only natural, but because my own changes didn't fit there. I was literally maladapted: my brain was the brain for a different place—my neuronal connections, my habits of mind, and ultimately my values had adapted to the woods. Just as I'd adapted to less sensory information, everyone else had adapted to more. I was like a tawny owl who had gone whiter in some remote snowy outpost, only to return to his native woods to find it had gone browner and now demanded brown to thrive. I'd anticipated having to adapt to being back in the city—in the woods there's nothing to filter out, nothing you don't want to see or hear; in the city you filter out most sights and sounds just to focus on what your friend is saying as you walk down the street—but there were also certain traits, certain rhythms of mind, that I didn't want to lose in the city. Not that I needed to remember every lyric from every aisle in the grocery store. But I wanted to hold on to a sense of place that included a sense of time, that was aware of the slow turning of the seasons and the stars spinning at night, a frame for my days that connected me to more than the daily, a kind of attention that could wait for answers to questions I didn't know how to ask, and a sense of identity that had more to do with how the seasons and starlight reached me, how they found a part of me I didn't choose, than with any flashes of light I could send out myself.

YOU DON'T NEED TO BE DARWIN to know that not adapting is a bad idea. In my tiny studio apartment near Fenway Park, I had a landline and email. I lived in the twenty-first century, or at least in the very, very late twentieth. But socially I might as well have been stuck in the nineteenth wearing my wild mountain-man beard and a coonskin cap. One woman, a talented Russian photographer, started our first date by saying, "So, no smartphone. You're making a political statement?"

She'd just gotten into the car. I tried to change the topic by asking about the restaurant we were going to.

"No, this interests me. Some kind of moral statement? About capitalism and advertising. About privacy?"

"I never thought of it that way."

She looked at me pityingly, the way you look at a dog that has just done something remarkably dumb and in his doggy-ness doesn't know it. There was a slight shimmer in her lipstick that caught the streetlight. She'd put time into doing her hair. As I pulled away from the curb, her disappointment seemed to expand beyond me to the dating pool in general.

I said not having a smartphone wasn't a statement, just a personal preference.

"You personally prefer not to enter the twenty-first century, why?"

I didn't know how to explain that I wasn't afraid of iPhones but of phone eyes, of having to adjust to yet another way of orienting in the world. I made an overdetermined joke about no one needing a writer to administer a semicolon in an emergency, received no response, then said something about loving airports as a kid, people always arriving from afar, each gate a portal to a different city, the feeling of the whole world brought closer together, but that as much as I still enjoyed passing through airports, I wouldn't want to feel as though I was living in one.

We had stopped at a traffic light. She turned to look at me. "Even homeless people have smartphones," she said.

NO QUESTION, NOT ADAPTING IS A BAD IDEA. A bad idea on the romantic level, a bad idea on the species level—just ask the dodo bird—and a bad idea on the neuronal level. Our brains evolved to have plasticity in order for us to survive, and evolution did such a good job that the basic structure of the human brain hasn't needed to change in the last forty thousand years. Our adaptations to new places and ways of life—to being hunters and gatherers in Africa, to living in agrarian communities in the Fertile Crescent, and to navigating the streets of modern-day cities (as a famous study of London cabbies has shown, the gray matter in their posterior hippocampi is notably greater than that of the average Londoner)—all happened within our forty-thousand-year-old brains, through the strengthening of links between neurons, through the growth of new neurons and synapses, and through the release of neurotransmitters. Basically, we begged out of evolution by developing brains adaptable enough to handle the challenges in our environments themselves.

This is the good news about our brains, but it may also start to be the bad news. For the majority of those forty thousand years, we had to adapt to only one environment at a time—whether on the savannah, in the forest, or on the farm. There was no possibility of a dual environment, of a magical man-made realm growing on top of the physical realm we walked through, no possibility of getting caught daily between two different environments, and between traits for one realm that were ill-suited to the other.

To be fair, this fear about environments, and the traits they might encourage, isn't new, even if early thinkers didn't use that framework. The first way, outside of daydreaming, to be in two places at once was through reading. Socrates derided this hot intellectual technology of his day as being little better than an intellectual Post-it Note. In the *Phaedrus*, written by Plato and not by Socrates himself, Socrates argues that writing would lead readers to have poor memory, an inflated sense of wisdom, and to be largely disagreeable, "since they will merely appear to be wise without really being so." He was right in a way. Reciters of long traditional epic poems, in India for instance, have been found to lose their astounding memory skills once they become literate. As readers, we don't have to "remember from the inside," which means we don't have to bring concepts into deep long-term memory, which is where we make connections with the concepts we've made part of our frameworks for understanding the world. And book knowledge is often decidedly far from wisdom. But Socrates was also wrong. He didn't anticipate the vast intellectual realms that reading might foster or how the traits it would help humans develop—analytical thinking, the capacity for sustained attention, and empathy for others—would be well-suited to the world beyond books.

But now we're effectively living in two places at once, the digital world and the physical world, hopping the border hundreds of times a day, and the cognitive traits we're adapting for one often hurt us in the other. The signs are everywhere. In our capacity for empathy (down 40 percent in college students over the past twenty years), in our capacity for political discourse (online incivility leads to polarization), in our attention spans (less than that of a goldfish), in our inability to be alone (six minutes without a phone and subjects chose to receive electroshocks rather than confront "solitude"), in our mental health (while correlation is not causation, suicide among American girls has doubled in the last fifteen years), and, quite literally, in our inability to be where we are (25 percent of US car accidents are tied to texting).

Meanwhile smartphones are only getting better at their discrete tasks—helping us to inform and be informed, to support and find support, to entertain and be entertained, to date, to navigate, to create, and even to meditate. So it's only getting harder to appreciate that they might also be getting better at functions they're not intended to have, like developing cognitive traits that might not serve us. They give us so much to see and wonder at, it might seem ungrateful to ask how they change *how* we see, even *how* we wonder. But just as the unintended consequences of fossil fuels are changing the earth's climate—causing certain species to thrive, others to suffer and adapt, and others to go extinct—the unintended, and sometimes intended, consequences of digital life are changing our cognitive climates—causing certain neuronal populations to thrive, others to suffer and adapt, and others to go, if not extinct, disturbingly dormant.

The catch is that going inside the brain isn't as easy as going to the Galapagos. The questions of what makes certain neuronal populations thrive and others suffer, and of how those neuronal populations become cognitive traits and ultimately values, are almost entirely hidden from view. (Currently, neuroscientists can monitor the activities of about one hundred individual neurons simultaneously, but a neuronal group consists of roughly one thousand neurons, and then there's the qualia problem, which is that we don't understand how neurophysiological events become the subjective experience of perceptions and feelings.) Once the technology catches up, no doubt scientists will study inner climate change more precisely. But for now, there's no official list of the endangered traits of our cognitive lives.

WHEN I WAS TWENTY, daydreaming out my dorm-room window, the question of what shaped the way I saw was so private I didn't even know how to ask it of myself. But now the notion that such a question would have bearing only on my personal life seems remarkably naïve. There's the danger, especially for writers, of mistaking the central question of your life for the central question of the culture, and that's what I hoped was happening during my sidewalk vigilante days. But since the 2016 presidential election, it has become sadly obvious that we're lost in a new way, disoriented in our very disorientation, no longer having any fixed stars beyond ourselves to find our bearings. You can't open a newspaper, or news app, without reading the latest jeremiad on how we know what we know or how we talk about what we talk about. Fake news. Questions about the very utility of facts. Our compasses are going crazy. We don't know where we are—so we don't know who we are or how to be.

Although I still don't have a cell phone, my window to the world turns more times a day than I'd like to admit into a computer screen. I now feel a quieter but more frequent version of that hard shift I felt all those years ago between the woods and the city, my brain caught again and again between two realms and their competing demands. And now I'm hardly alone—in the US people swipe or tap an average of 2,617 times a day, the jumping back and forth so frequent the resulting disorientation has blurred into a nameless unease. We're forgetting we're lost, and we're forgetting what we're losing, which is a far cry from being well adapted.

We need a new kind of map. A map with the digital world and the traits it calls for, and with the old physical world and the traits it calls for, and with the borders clearly marked where the two realms conflict—where the border crossings are treacherous, where we're bound to lose parts of ourselves we value.

This book is my attempt to draw that map. If I were a visual artist, I'd have designed one like those at the beginning of fantasy novels, complete with mouth-bending names, alluring reaches, crowded ports, surprising wastes, and daunting mountains passes of little triangles. But stick figures are my upper bound, so I've simply used the endangered traits I fear losing most as a way to draw the borderlands. It's a personal map, of course—it won't correspond entirely to your experience, or to the parts of yourself you value most and want to protect.

But it can be, I hope, a start to charting your own map. A few more stars, not in your pocket, to steer by.

CHAPTER 1

A Map to Our Maps

ENDANGERED TRAITS

Lostness (from Old English *losian*, to perish, to lose, akin to Old Norse *losa*, to loosen, to Latin *luere*, to release, atone for)

Memory palace (Memory from Latin *memoria*, from *memor*, mindful, akin to Latin, *mora*, delay; and Palace from Latin, *palatium*, from *Palatium*, the Palatine Hill in Rome where the emperors' residences were built)

Survey map (Survey from French *surveeir*, to look over, from *sur-* + *veeir* to see; and Map from Latin *mappa*, napkin, towel)

BACKGROUND

My grandmother died at the age of 101, having lived the majority of her life in Newburgh, a small city about an hour up the Hudson River from New York City. After her funeral, my parents and I drove from the cemetery to a lunch reception at an Italian restaurant overlooking the river. Newburgh always makes my parents nostalgic—they grew up there, were high school sweethearts—but the extra space in the sky this late June morning, the years floating down the streets and drifting above the familiar buildings, seemed to pull them through the narrow straits of nostalgia into a wider kind of wonder. Perhaps all the arrivals my grandmother had enjoyed—cars, television, moon landings—and all the departures she'd suffered—the early death of her first husband, the relatively early death of her second husband, and the deaths of all her siblings—had pushed open an additional sense of awe in my parents at the span of their own lives, at the unlikely distances they'd traveled.

Dad decided to take a slight detour. Broadway, Newburgh's handsome and thriving center of commerce in the '50s, still slopes grandly towards the Hudson, brick storefronts lining the way. We passed the boarded-up windows, the payday-loan centers, the storefront churches and liquor stores—the avenue's expansiveness a showcase for all that is no longer there, for the hard times that have hit inner cities across the country. Mom pointed out where her grandparents' clothing store had been; she'd worked there after school. Dad pointed to the old *shul*. Together they placed Broadway's three movie theaters. "Only one would show *Tarzan*," Dad said. Then he exclaimed, "Look, Pete's Hot Dogs. Still open for business. Best hot dog in Newburgh!"

"Probably the only hot dog in Newburgh," Mom said.

She pointed out the seamstress shop, asked if they'd told me the story of how she had sewn her apron to her skirt in Home Ec; Dad pointed out the bar that would let him shoot pool upstairs after school; and they kept marveling at how hard it was to explain how it looked, this broad avenue with the view of the river, how it really had been the heart of Newburgh, which really had been designated "an All-American city."

When we got to the restaurant, my niece Sophie, a reedy eleven-year-old, tugged on my hand and asked where we had been. "Did you get lost?"

"We took the scenic route."

"I don't believe you. You got lost."

"Have long have you been here?"

"Like ten minutes! You got lost. Admit it."

I overheard my sister-in-law, with slightly less persistence, asking my mom the same question. "You didn't get lost, did you?" She prides herself on her navigation skills—my brother's nickname for her is Mapgirl—and they always arrive before my parents, no matter the destination or occasion, then inquire as to my parents' route. It's become a joke between my parents: How long until the kids ask how we got here?

As lunch wore on, my niece kept circling back, kept pressing me to admit the truth. When I tried to explain that her grandparents had taken Broadway not because they'd gotten lost, or because they'd thought it would be the faster route, but because they'd wanted to drive down a street that stirred memories, that reminded them of who they'd been when her great-grandmother was younger, she eyed me sidelong. "You're just making excuses." Then she brought her face close to mine and mouthed the words, as though they were too shameful to say aloud: "You got lost!"

I felt a little lost, between her sense of the streets outside the restaurant and the sense of those streets my parents had given me, and I was at a loss as to how to unify the two, how to slide them back into one city. It was a minor vertigo, an invisible collision between the streets as holders of memory and the streets as a test of efficiency, as though someone kept shifting the orientation of the room.

EARLIER GENERATIONS WERE more attuned to invisible collisions. On a cold clear January night in Padua in 1610, Galileo Galilei peered through his telescope in the garden behind his house and saw three tiny specks of light around Jupiter. The next night he set up again and observed that the three tiny specks had moved, and a few nights later, he set up again: there weren't three tiny specks moving but four. His observations contradicted the geocentric belief that celestial bodies only orbited Earth. Plus, the geocentrists had claimed Earth couldn't possibly move through space—if it did, they argued, the Moon would "get lost" and wouldn't be able to keep its orbit around Earth. But Galileo observed that the moons of Jupiter clearly had no problem staying with Jupiter. All of which meant Copernicus's heliocentric model was correct or, at the very least, that the arguments against it were wrong.

Pressured by a range of Church officials to pipe down, Galileo waited twenty-two years, until he was sixty-nine years old, to publish his findings. The book was written for an educated public, couched as a debate among three gentlemen, and the printing sold out. That summer, after a February publication, the book was banned. On the first of October, Galileo opened his door to find the Inquisitor of Florence issuing him a summons. The following June, seven of ten cardinals in Rome signed a sentence formally condemning Galileo, demanding that he recant his findings, and calling for the house arrest he was to live under until his death. It took another October morning, and another 350 years, for the Church to issue a formal statement saying that Galileo was right.

The outline of the story suggests that the Church was hostile to science, but that wasn't true. Pope Urban VIII often held eager audiences with Galileo; the pope's private secretary urged Galileo to publish his ideas. What the Church was hostile to, once the ramifications of Galileo's ideas became clear, was that his new map of the cosmos was also a new ethos of place. For the Church, a map with Earth at center of the universe meant a theological map that made sense: God had created the universe for Man, which meant we were the center of God's creation, and the stewards of it, and the Church was the steward of the stewards. Good map, good ethos, good hierarchy. But a map with the Sun at the center of the universe meant a theological map that no one knew how to read. It contradicted what anyone tracking the sun moving across the sky could see; it contradicted the long-accepted theories of Ptolemy and the teachings of the Bible; and it threw into question not only the relationship between Man and God and the rest of creation but also the relationship between the Church and science, effectively remapping who got to shape our sense of our place in the universe.

In other words, the Church recognized a map of the solar system wasn't just a map of the solar system. It was a value system.

I DIDN'T REALLY GET THE JOKE AS A KID, but I loved Saul Steinberg's poster "A New Yorker's View of the World." It hung in the dining room of my friend Ronnie's house and promised vistas of insight beyond any maps in school. The poster looks west across the country and shows, starting at the bottom, Ninth Avenue with pedestrians and parked cars and traffic lights, then Tenth Avenue in only slightly less detail, then the Hudson River with a few docks, New Jersey as a bland strip of land, Chicago as a word hovering in the featureless void of the Midwest, then a flat nameless expanse stretching west to the Pacific, the humps of China, Russia, and Japan like three breaching whales in the distance.

Nowadays, thanks to GPS and its genuinely groundbreaking orientation, we all get to be New Yorkers: we're all back at the center of the universe. The top of the maps in our cars and on our phones doesn't face north, or even east (as some ancient maps did, orienting towards the sun); they face whatever direction we're facing. Each of us is the center of the map, as the map moves with us, and each of us is the orientation of the map, as it literally revolves around us as we change direction. No more geocentric or heliocentric view of the universe—we've innovated a map to match our psychological default setting: egocentric! However physically and logically impossible the egocentric model, it's psychologically pleasing and neurologically efficient. We no longer need to know which direction we're facing to get where we're going. We no longer need to pay attention to the direction of the sun, that fallen god. Rather than orienting by landmark or memory, we've trained the streets to orient, and orbit, around us.

Also in the name of efficiency, most GPS maps have no need for geographical data other than the roadways and their names.

Waze, for instance, replaces the timeless features of rivers and mountains with useful "real-time" updates on car wrecks and speed traps—its only temporal dimension an estimated time of arrival, the millennia it took the mountains and rivers to form quietly flattened into the minutes of a daily commute. Which means we're the center of the universe not just geographically but also temporally, which is even more self-centered than Ptolemy, who at least had the cosmos created before our existence. But now our maps don't have time for time. If a feature of the land doesn't pertain to our efficiency, the assumption is it no longer pertains to us.

WHY IT MATTERS

On a trip not long before my grandmother died, my brother and his wife got in a fight. My brother had just bought a new GPS—he'd named it Martha—and he wanted to use it on the drive from Newburgh into Manhattan. His wife, a.k.a. Mapgirl, didn't see the need. She rarely gets lost in new places, and Manhattan wasn't new to her.

"Would just you turn it on?" he said.

"Why don't you trust me?"

"I trust you. But this is why we got it. This is Martha's purpose."

My sister-in-law turned on the GPS, her unsaid words likely caught in a traffic jam, rerouting, rerouting. In the backseat I was pretending not to exist. This was Martha's purpose. But what was her purpose, I could almost hear my sister-in-law wondering. What about Mapgirl?

We followed Martha's crisp directions out of Newburgh. I imagined Dante the Pilgrim in the dark wood with a GPS: no more losing his way, no more soul-searching (literal or metaphorical) through concentric circles of sinners—and so, ultimately, no redemptive trip into Paradise. I imagined Odysseus with a GPS on the wine-dark sea: no more losing his way back to Ithaka, no need to abandon a sex-thirsty nymph, no need to outwit a blood-thirsty Cyclops—and so, ultimately, no need to remember the love and duties that bound him home, and no heartbreaking homecoming. At most, Dante and Odysseus would have suffered road rage (or sea rage at Neptune and his storms), indignation at a few wrong turns, and frustration with an incorrect estimated time of arrival, with trying to figure out how much to blame the GPS, how much to blame themselves, and how much to blame fate. Without getting well and truly lost, Dante and Odysseus wouldn't have had to learn how to be lost, and their outward searching wouldn't have turned inward.

Not that I wanted to get lost in the swamps of Jersey with my brother and sister-in-law. I was meeting two friends on the lower West Side for brunch, which didn't leave much time for escaping the cave of a Cyclops. Besides, we were making good time. But I still felt an uneasy tension between Martha's purpose and Mapgirl's purpose, and more broadly between Martha's purpose and something primal and spiritual in all of us—the ability to find our way, which includes at times the ability to get lost.

A thin November snow had begun spitting at the windshield, and I remembered how once in a late January storm, in the hills near the house in Vermont, the familiar path back to the meadow had vanished: the woods a spinning whiteness, the outlines of bare trees like the ruins of a long-lost alphabet, as though I'd been dropped on a long-uninhabited part of the planet. I couldn't visualize myself in relation to the house, couldn't find any map in my head that corresponded to what I was seeing. The shape-shifting hills looked vaster and stranger than ever, a haunting place humans perhaps weren't supposed to be, which only made me feel more human, more aware of my need for shelter and fire. Eventually I hit a snow-covered dirt road, and the land fell back into a place, a map blossoming in my mind: the house, the arc I must have traveled through the woods, the road down to town. But the map wasn't as solid as it had been—the woods surrounding the house felt bigger, older and more mysterious, the hills having flashed a reminder of themselves without the superimposition of any map at all.

In the front seat, my brother and his wife had begun talking again quietly. We would arrive on time, just as planned. But as Martha's cool steady voice zipped us by the strip malls on the Palisades, I couldn't help feeling diminished, like we were bad human beings—not in the moral sense, but bad human beings in the sense of being bad animals. Attention had become decorative rather than essential. There was no need for us to consider anything outside the window, no need for us to remember any landmarks or history, and no possibility of mystery, of the unmapped or the unmappable. We'd been freed up for more important activities, which, from the conversation in the front seat, was apparently the highway version of window-shopping—considering what we might order online once we were back in our respective homes. The land was just a series of efficient corridors we were passing through, linking us from one particular destination to another, without forcing us to confront all that was outside those corridors, all that connected them.

AS A BOY, I always wanted the map in my hands. Not to question the route—long car trips were almost always between Boston and Newburgh, and my parents knew the way as instinctively as migratory birds—but to track our progress by watching the signs change along the highway, finding where we were on the map, and then watching the signs change again. This gave me immense pleasure sitting in the backseat, little legs folded Indian-style, map spread to the edge of the territory allowed by my brother. Sometimes I'd anticipate the upcoming signs, sometimes I'd take an exit and follow the thin curving line on the map, my finger arriving with great excitement in Sturbridge or Litchfield. From there I'd branch outwards, imagining how long it might take to journey somewhere exotic like Schenectady or Buffalo: the land kept on going, and the map kept on going, and if we kept on traveling on the land, then we could reach the place on the map—and we could see what was there! These solo mental excursions felt like covert training for real grown-up adventure.

But becoming a young man, it turned out, had far more to do with learning how to be lost. The summer I was twenty-four, I inadvertently made my first foray into solitude. An ex-girlfriend, whom I'd fallen in love with during a year in Italy, invited me to spend the summer at her family's farmhouse in rural Austria. She'd be in Vienna, a two-hour trip by train. The house was for sale, emptied of furniture other than a mattress, a desk, and a small working kitchen on the third floor. I pictured weekend visits, long strolls into the mountains, long talks into the night. I pictured lounging with her in the garden of her childhood stories, impromptu weekend road trips to Italy and to France. I pictured the distance between our lives collapsing, circumstance a doddering chaperone we'd escape, only to find ourselves undeniably back in the rush of our months together in Italy, undeniably back in love.

But she didn't come to visit. She simply said, again and again, she had too much work. She'd hang up, I'd open the phone booth, and there I was: still in the center of a little farm town in Austria, knowing only a few phrases of German that seemed more useless and more apt, phrases from Rilke about solitude and sky. When I bicycled to the one bank in town to sign traveler's checks, I was afraid my signature might have changed due to a broken heart and that my money for the summer would be worthless, my identity no longer clear. But the teller-in-training, after consulting with her manager, handed me back some schillings, and I bicycled back to the empty house, and later that afternoon rode several miles along a stream to a lake my ex-girlfriend had once mentioned, and soon found myself in the slow evenings that descended, one after the next, listening from the window on the third floor to the different sized bells identifying the different cows on the neighboring hillside, and taking long after-dinner walks along the brook that ran down the hill through the pines, aware, as summer deepened and the garden grew wild, how little I knew about where I was, about the town and the region and its history, and how little I knew about what I would do with my life once I returned to the States. But during those bike rides through her country's fields, the train passing with passengers whose lives I couldn't imagine, during those late-night walks surrounded by a darkness that wasn't mine, I slowly came to feel comfortable with being lost. The answers I was looking for, the questions of who I would be, and who I would love, and how I might eventually fit into a place, would all take time. But figuring out who I was in between the cracks of all those answers, in between the cracks of place and time, suddenly seemed important, a kind of preparation for all the questions that would follow, even though it was itself a question I'd never thought to ask.

If I'd had a GPS with me that summer, maybe I'd have found a quicker route to the lake, maybe I'd have found another nearby town to bicycle to. But if I'd had a GPS mindset, those two months would have been a lost summer. Not a summer of lostness, which is a very different thing.

THREATS

The default scale on GPS maps allows us to see the segments of our route as we proceed, with no extraneous information to distract us. While this might be efficient in the short term, the result is that we gain only *route knowledge*. Harvard physicist John Huth explains in his book *The Lost Art of Finding Our Way*: "If we have route knowledge, we know routes and landmarks along these routes. We know of a network where different routes join each other but not what lies between the routes themselves." The alternative to *route knowledge*, Huth explains, is *survey knowledge*. "Survey knowledge is a complete familiarity with an environment. In your mind you see the region as if you are hovering over the landscape and seeing everything below in miniature." Given our ancestors' need to find their way to the watering hole, or back home to the cave while avoiding roving predators, their brains evolved to develop survey maps, as did the brains of all mammals. That way they could detour around the creature with the flashing teeth and the fearsome roar, even if it didn't yet have a name.

The psychologist Edward C. Tolman first discovered the brain's propensity for route maps in 1948. He put rats in a maze, gave them time to find the convoluted path to the food, then replaced the initial maze with a second that was similar but had possible shortcuts, and the rats took the shortcuts. Tolman concluded that the rats had figured out where the food was relative to their starting position—they'd developed survey maps, not route maps.

You could argue that if today's rat knew how to read a GPS, he'd get the latest shortcut updates and find the food just as quickly, so what's the difference? To a rat there might not be one. But to a person the difference would depend on whether he was solely interested in finding the food. Route maps efface the places in between our starting point and our destination into flyover country, a series of nowheres to be endured as a means to an end. According to the philosopher Immanuel Kant, one of the fundamental principles of morality is, "Act in such a way that you treat humanity, whether in your own person or in the person of any other, never merely as a means to an end, but always at the same time as an end." It's arguable whether there can be a comparable morality of place, but to treat everywhere but your destination as a means (which Martha's cool imperatives encourage) relegates the in-between places on the map to second-class status, and, more importantly, may relegate the in-between places in your own psyche—the curiosity, the sense of adventure, the desire to orient yourself as a part of something larger than yourself—to second-class status as well, constantly rerouting you back towards your original destination, towards efficiency, towards no tolerance for error, which means no tolerance for wandering. By contrast, if you've developed a survey map of where you are and where you're going, you can afford a few wrong turns, still have a basic sense of orientation, and not feel lost but capable of exploration. "I never was lost in the woods in my whole life," Daniel Boone said, "though once I was confused for three days." Not that anyone would want to be confused for three days. But for Boone being confused just meant he had to pay closer attention to the sun, the stars, and the trees, and had to work harder to navigate by what he knew and by what he could observe. Even for the less navigationally adept, curiosity, heightened attention, and humility in the face of the unknown have always been part of traveling. But now we seem to be losing the human orientation of orientation and disorientation—never really getting lost, geographically or in the kinds of thoughts and feelings places can inspire, and never really finding our way.

A friend, a French professor who grew up in the Bronx, recently returned from a trip to Italy. The city she'd loved most was Venice. When she'd asked her hotelier for a map, he'd said, "There is no good map of Venice. Go and get lost. That is the pleasure."

ONE OF THE FRIENDS I was going to see on that trip to Manhattan was Oliver Sacks. We'd met a few years earlier at an artist residency in the Adirondacks, where we had neighboring studios. In the evenings, on our long shared porch, Oliver and his partner, Billy, would invite me to share a bottle of wine, and we'd talk as the lake darkened through the trees. Oliver's questions weren't those of a doctor, and they weren't quite those of a new friend. He had lost vision in his right eye a few years earlier. Strangely, Billy's father had only been sighted in one eye too. Oliver asked things no one had ever asked me. Did I dream with depth perception? When Oliver did, he said, he awoke with a profound sense of loss. Did I fear losing vision in my left eye? His fear, he said, sometimes overwhelmed him. Did I dread rainy days, with their threat of umbrella spokes? He certainly did.

I'd been going to visit him and Billy in the West Village about once a year ever since, always amazed by how much wider the streets looked, how much more wondrous and mysterious the people, after spending an hour or two with them over tea. One afternoon, prompted by something I said about his teapot, Oliver remembered a patient for whom the perception of time intermittently froze, the tea pouring out of her teapot resembling an icicle, until time unfroze and she promptly saw that she'd poured tea all over her table. This in turn led him to recall a patient whose visual perception froze while her auditory perception remained intact: at a busy intersection, she would encounter what appeared to be stopped traffic, though she could still hear the wind-rush of passing cars. She had to learn to trust her ears and not her eyes—to listen for what was really happening—something I'd had to do after my eye accident. To say goodbye to Oliver and Billy, take the elevator downstairs, and walk back outside into a Sunday afternoon on Horatio Street was to emerge to a beautifully complex world: a passing cyclist, and what it took for her to perceive the shifting traffic patterns, to maintain balance as a pedestrian stepped off the curb into her way, and for her to be singing along to the song in her ears, recalling the lyrics effortlessly, as she shot past; and the guy stepping forward into the street turning to look after her, then pulling his dog along on a leash, cascades of information pouring through his brain and through the dog's brain, and yet they crossed the street so easily, gracefully even, and standing on the sidewalk I could see them across the way, could hear the rhythmic jingle of the dog's tags above the distant roar of city traffic, could consider their progress down the block, all while feeling the late afternoon sunlight on my skin.

I DIDN'T MENTION my GPS experience with my brother and Mapgirl, but I'm sure Oliver and Billy would have appreciated it. Only two years later, keenly aware that cancer wouldn't grant him much more time to live, Oliver wrote, "Over the last few days I have been able to see my life as from a great altitude, as a sort of landscape, and with a deepening sense of the connection of all its parts. This does not mean that I am finished with life. On the contrary, I feel intensely alive, and I want and hope in the time that remains to deepen my friendships, to say farewell to those I love, to write more, to travel if I have the strength, to achieve new levels of understanding and insight." It's a passage I find almost unbearably beautiful—his grace, his enduring love of life, and his perspective, which he imagines as a kind of survey map of his life, with no part of it as flyover country. No part of his life does he need to distance himself from. Not his mother's brutal condemnation of his sexuality, not his years of dangerous experimentation with drugs, not any of his professional disappointments. Oliver didn't know how much time he had left or how much pain he'd have to endure, and he wasn't without fear, but you can feel in his words a profound patience with the ultimate lostness, his sense of being well-enough oriented in his life to be comfortable with the greatest unknown.

IF WE NO LONGER develop survey maps of the places we live and travel through, will we still be able to develop survey maps of our lives? Will we still be able to achieve a kind of orientation that is really a kind of wisdom?

The concern isn't as metaphorical as it might sound. Neuroscientists have found that when you picture how to find your way—whether you start from a map or the memory of your previous trips and the internal map you formed from them—you rely on a part of the brain called the hippocampus. That famous study I mentioned earlier of London taxi drivers revealed that the drivers had more gray matter in the posterior hippocampus than the average Londoner. From years of navigating London's labyrinthine streets, their brains had changed: as the hippocampus did more cognitive mapping, it became bigger. Inside the brains of the taxi drivers, the hippocampus developed like the quadriceps of a runner—enhanced from regular workouts and ready to carry a heavier load.

A fit hippocampus isn't the easiest thing to show off at the beach, but, sadly, atrophy of the hippocampus might be. Several studies have linked such atrophy to increased risk for a range of psychiatric disorders, especially dementia. Whether relying on GPS year after year might contribute to that degree of atrophy isn't clear, but studies have shown that GPS users are more likely to suffer from problems with memory and spatial orientation. Veronique Bohbot, a McGill University neuroscientist, has said, "We can only draw an inference. But there's a logical conclusion that people could increase their risk of atrophy if they stop paying attention to where they are and where they go."

Since the ancient Greeks, we've known that remembering place helps us to remember far more than place. Mnemonists have long used a method called the memory palace to store arbitrary lists of information. In his book *The Mind of a Mnemonist*, the neurologist Alexander Luria describes how his subject S. was able to remember every item on random lists years after memorizing them because S. "would 'distribute' them along some roadway or street he visualized in his mind. Sometimes this was a street in his hometown, which would also include the yard attached to the house he had lived in as a child and which he recalled vividly." Tracing his path along this familiar route enabled S. to remember every piece of information as though it was a necessary part of finding his way. He remembered the random information because his hippocampus assumed he needed to remember it in order to navigate. This process suggests not just a kind of time travel in personal memory—to S.'s childhood—but also a kind of time travel in human history, back to pre-linguistic times when new information would never arrive in the hippocampus divorced from the context of place, from the path to the watering hole or the hunting grounds, or divorced from the urgency of survival. We had memory long before we had language, and it was inextricable from place.

But by not using the hippocampus as we navigate with GPS, we're interfering with the brain's own natural multitasking, its ability to consolidate memories through an awareness of place. We're interfering with a cognitive ecosystem that we don't fully understand, especially when it comes to how important our sense of place is to how we remember and make sense of our lives. We may save time with GPS, but we're likely not keeping it, not in any meaningful way. Myopia towards our experience of place, in our maps and in our lives, becomes a kind of myopia towards our experience of time.

A map isn't just a map. However inadvertently, it's a value system.

WHAT YOU CAN DO

This past Thanksgiving, for the first time in more than twenty years, my parents and I didn't go to Newburgh. With my grandmother no longer alive, we didn't make the trip. We shared the holiday meal at my cousin's house in Connecticut, and then my uncle on my dad's side drove up from Newburgh on Friday and met us "halfway," at a diner in a town none of us knew. It was more convenient, saved us time in the car—a blessing—but it also made me feel a little disoriented. Every return to Newburgh is a return to my parents' stories, to hearing again about the house my mom's father built, the convenience store he'd walk to every Sunday morning for the *New York Times*, the front step where she fell and banged her head (a small bump still there on her forehead), the intersection where she and her sister would sled and their dad would block traffic, the neighbor's house whose door was always open to her for a BLT, the house where my dad's best friend, Bobby Hegel, who died young in a plane crash, had lived as a boy, the JCC where they used to shoot pool, the YMCA where my dad's father, who was sickly and died the month before I was born, used to swim, the junior high ballfield where my mom and dad played Red Rover and she called him over, the school gym where he asked her what she'd say if he asked her to go steady, the water tower where they went, to sit and talk and kiss once he finally did, the high school where he cheated off her in Spanish class, the house where she threw him a sweet sixteen party, and the pond where she skated off without him, upset about something she no longer remembers, leaving him to crawl along the ice. I even missed our visit to the cemetery, where every year we'd put pebbles on the headstones of my mom's parents. After she'd dried her eyes, Mom would point out how the whole town and its history were there, the dentists and the teachers and the secretaries, the love affairs and the feuds, all of it still present in who had wanted to be buried next to whom.

Part of my sadness at not visiting was the fear that I'd never get a chance to hear their stories again, that we'd never again have time or reason to make a full tour through Newburgh. Maybe this was partly the fear of losing my parents, but they were both in their early seventies and in good health. Maybe it was partly the fear of losing Newburgh itself, a town that was always somewhat lost to me as I didn't grow up there, and became important as a kind of living family photo album whose stories we could drive or walk through on every visit. But it was also the fear of losing my mom and dad's way of experiencing Newburgh—of knowing place through time, and of knowing time through place, and of knowing who you are and how you're a part of the world around you through the inseparability of the two.

As we swung onto the highway back towards Boston, our ETA clear on the GPS, I wondered if I was just being nostalgic for my parents' nostalgia. Or even for our routine. There was no visit to the cemetery, no stopping at the Sunoco where we always gassed up, no Beacon Bridge and sweeping view of the Hudson. Instead, there were unfamiliar signs along the highway, no need to look at them because of the steadily advancing blue line on the GPS. The bare nameless land slid by outside the window, the placeless rivers, the placeless hills. Here and there a Target. A Wendy's. A Taco Bell. The land felt both foreign and smaller, not known but with no need to be known.

"That worked out well," said Mom half-heartedly.

"We'll make good time," said Dad.

It felt like we'd lost the map to our map, like we knew how to get back to Boston but didn't quite know what we were steering by.

You got lost, I heard my niece's voice say.

CHAPTER 2

Clocksetters

ENDANGERED TRAITS

Event time (Event from Latin *eventus*, from *evenire*, to happen or come out from; and Time from Old English *tima*, akin to Old English *tid*, tide)

Flow (from Middle English, *flowen*, akin to Old High German, *flouwen*, and to Latin, *pluere*, to rain)

BACKGROUND

You don't need a clock to know what time it is—you don't even need to follow the arc of the sun. Time's reflection is everywhere. I first noticed this during my years in the Vermont woods as an absence. Without registering it, I'd told time in college by flock migration: the stairwell scurry in the dorm to catch breakfast; the scraping back of chairs in the dining hall to bus trays; the packing up of backpacks roughly sixty seconds before the professor stopped lecturing. But without any human reflection of time, the Vermont woods felt timeless at first. Not measured out in coffee spoons but in full moons and deepening nights, measured by the movement of the Earth relative to the Sun, and of the Moon relative to the Earth, measured by the kind of time that existed before we did, which perhaps is why we call it timeless. But soon I became attuned to the two- and three-note morning scales of the chickadees, and the sharpening scents in the meadow as the sun dried the dew on the high grass, and the crows towing dusk behind them as they returned to their roosts beyond the pines. Those glimpses of time, along with the passage of the sun itself and the passage of Orion in the window above my head at night, formed my new clock in the woods—a clock that often allowed me to forget time, to enter pockets of the day in which time didn't seem to be passing at all.

In his book *More Scenes from the Rural Life*, Verlyn Klinkenborg writes about the inverse phenomenon, about becoming a clock for the natural world—or, at least, for his farm animals. One October afternoon, on his way to the mailbox, he catches them lounging in their animal lives, utterly unaware of him and utterly unaware of time. The horses are standing broadside to the sun, sunning their flanks, "basking ... in the long, floating, unscheduled middle part of the day," the hens are primping and pecking in the dirt with no concern other than primping and pecking. But Klinkenborg's presence signals to them a feeding schedule about to be enacted, their needs about to be met.

> Whenever my life intersects with theirs, it's all expectation, a concern for what comes next. Among the animals, I feel as though I carry time like a bacillus, and we share the infection. But in the middle of the day, equidistant from morning and evening chores, the animals drift away into a life all their own. But if I stepped into the pasture, hoping to still my own clock, to pause in the sun, the horses would wander over—and the chickens would rush my way—as if to ask, "What now?"

Klinkenborg regrets that he can't avoid being a kind of clock-setter to the animals, and also that he can't slow his own clock to their pace, "to pause in the sun" and drift into a life all his own. But setting Klinkenborg's clock are the feeding and tending duties of his small farm and, even from a distance, the pace of the outside world, with its demands that mail be opened and answered in a "timely" fashion. His description of life on the farm makes clear that every creature, horses and chickens and humans included, lives inside a time framework, one highly susceptible to change but not to its own control.

Something or someone is always setting our clocks. And we can get caught between clocksetters, caught in a jet lag not of time zones but of kinds of time.

PERHAPS THE TWO BIGGEST technological influences on our sense of time, until recently, were the railroad and the factory. Before the railroad, high noon whether you were in Omaha or Oakland was when the sun was highest in the sky. Each town had its own timekeeper—the official time generally kept on the town's church clock or city hall—which allowed each town, to some degree, to have its own pace. But once the railroad connected Omaha and Oakland in 1869, and a very long string of towns in between, all those idiosyncratic noons presented major scheduling and safety problems. There were innumerable missed trains. There were numerable crashes. The pocket watches of stationmasters all along the line needed to agree with each other, and passengers needed not to be late. So in 1883 the railroad adopted "railroad time," instituting time zones and standardized time across the country, and most towns set their clocks to match. But not everyone liked the change. In Cincinnati, where the clocks would lose twenty-two minutes to agree with the Pittsburgh clocks, a writer for the *Commercial Gazette* wrote, "Let the people of Cincinnati stick to the truth as it is written by the sun, moon, and stars." In Boston, the *Evening Transcript* cried, "Let us keep our own noon!" In Indianapolis, one reporter for the *Daily Sentinel* worried that people would now have to "eat sleep work . . . and marry by railroad time." The entire city of Detroit refused to abandon solar time and turn back their clocks twenty-eight minutes to Central Standard. But in 1918 the federal government standardized the four time zones by federal law, and everyone had to get on board.

Around the same time, a man named Frederick Winslow Taylor did something revolutionary: he brought a stopwatch into the Midvale Steel Works in Philadelphia. He'd noticed that the laborers on the plant floor each worked at their own pace, some more industrious, some more leisurely. Perhaps one man liked to hum "My Darling Clementine," a pop hit at the time, as he worked; perhaps another liked to work fast, then pause to enjoy a cigarette. These individual paces led to gaps in efficiency, gaps in productivity—each man's personal sense of time not fitting together as smoothly as the parts they were assembling. So Taylor recorded the speed of the workers and the machines, broke down every action into discrete steps, and analyzed the data for efficiency. He brought science, as he put it, into the workplace and made the men more machine-like. Speed became the almighty coefficient for productivity. This may seem obvious to us now, but many workers and managers at the time weren't long off the farm, and in the fields there were only so many heads of lettuce to pick—you could go faster or slower, so long as you finished the day's job before nightfall. But in factories, assuming the supply of materials didn't dry up, there was no limit to how much you could produce, other than the human limits of endurance (which Taylor tested in more ways than one). The workers didn't like him timing how long it took to sit or stand or open a drawer, claiming that he was making "every man merely a cog or a nut or a pin in a big machine." But management loved him. He had commodified time.

By 1907 workers around the country were dutifully punching a time clock on arrival and departure. The dominant player in the punch-clock business was the International Time Recording Company, which later became IBM.

ROUGHLY A CENTURY LATER, my heraldic ENVY 5660 printer informs me of a fundamental change in the human experience: "Now Time and Location are irrelevant." The silver words float across my screen during installation, blazoned on a banner wavering in a digital breeze. All that's missing is a trumpet fanfare. I can do whatever I want (at least printing-job-wise) from wherever I want whenever I want. Perhaps if my friends knew about my printing capabilities, a wildfire of envy—are there really 5660 kinds of envy?—would catch on the directionless breeze.

But there's a hitch to my new freedom. Instead of just making time and location obsolete, Taylor's stopwatch, with its new superpowers of efficiency, has introduced a new kind of timeless time. There's no more simultaneity. I'm not talking about my printer now but about my Facebook and email accounts, and of course about texting, and how we're all checking for a message that hasn't yet come, or getting a message too late, or sending a message not knowing when it will be received. This new time is less railroad or factory time than a cross between Sartre's *No Exit* and an Oscar Wilde farce. In *No Exit*, the three characters are stuck in a room together, unable to leave or even turn off the light, stuck in "life without a break." Inez says, "To forget about the others? How utterly absurd! I feel you there, in every pore. Your silence clamours in my ears. . . . It's all very well skulking on your sofa, but you're everywhere, and every sound comes to me soiled because you've intercepted it on its way." But now we're stuck with this constant awareness of others, and of ourselves in the eyes of others, without actually being in the same room; so texting or posting or emailing back and forth is like a Wilde character entering a room, finding no one, leaving a note on the table, exiting, returning anxiously to find no response, exiting again, another character entering through a different door, writing a response to the note and exiting, the first character returning breathlessly, wondering how he missed the second character, reading the response and recognizing his own message has been badly misinterpreted, trying to clarify, exiting, whereupon a third character enters the room . . . etc. Watch anyone, especially a teenager, checking her phone, putting it facedown on the table, checking it again, typing another message, putting it down closer to her, picking it up, finding nothing, putting it down faceup, then needing to hold it, needing to hold the entrances and exits in her hands, and you're watching someone caught in this new kind of time. Meanwhile the possibility of being in the same moment as the person sitting across from you, who is likely also on his phone, has vanished too. We're all trapped with each other outside the present moment. *No Exit* is Sartre's version of Hell.

Which may hardly sound like Taylor's dream of efficiency. But for companies like Facebook and Twitter and Yahoo and Google, it is highly efficient. They know what our attention is worth down to the second, the click, the penny—which is how they determine their enormous bottom lines. In other words, Taylor's emphasis on efficiency spread from the factory to everywhere not just because the office is now everywhere but because, in effect, the factory is now everywhere. Silicon Valley wants to extract the most from us, its unwitting laborers, and their applications need our eyes, our clicks. They need our free work so they can sell our data to advertisers. Our time is their money. And so they make our attention efficient for them, seducing us to click as much as possible, while luring us, ironically, with the promise of efficiency for ourselves. They tell us that efficiency—whether for business, or our social lives, or anything else—is the new standardized time. Time zones, let alone the sun and the stars, hardly matter.

Meanwhile the factory whistle keeps calling us in evermore seductive ways. We keep running to check the empty room. And just like that, we're on the clock.

WHY IT MATTERS

My parents have had the same fight for over fifty years. Imagine a late drowsy Sunday afternoon, a woman, let's make her middle-aged, lounging on the couch with a book, a man in a wool sweater snacking on peanuts in the kitchen. The woman looks up. "What time did Michael say?"

The man emerges from the kitchen, offers a handful of peanuts. She shakes her head. "Seven thirty," the man says. "But you know how parking's impossible. We'll need to factor in fifteen minutes to find a spot, another fifteen to walk to the restaurant. Plus twenty minutes for the drive, so we should be in the car downstairs ready to go at six thirty, just to be safe. Are you showering? Now if you're going to shower, you should—"

"Don't!"

"Honey, I'm just telling you—"

"I asked what time Michael said."

"I know what you asked. And I know what always happens."

"You want to be in the car at six thirty. Thank you. That's all I need to know."

She would turn to me. "He makes me crazy. A Sunday afternoon chopped into little bits and gone."

He'd shake the peanuts in his hand like dice. "If you want to get there on time, and you want to shower—"

"Three words. Im. Poss. Ibble."

My bet is they had roughly the same conversation before their prom, albeit less directly, and before their dinner dates in college. I know they had it before my bar mitzvah, and before my high school graduation, and before my college graduation. Mom has always been more comfortable living on event time; Dad has always been more comfortable living on clock time.

"One of the most significant differences in pace of life," writes the social psychologist Robert Levine, "is whether people use the hour on the clock to schedule the beginning and ending of activities, or whether the activities are allowed to transpire according to their own spontaneous schedule. These two approaches are known respectively as living by clock time and living by event time." Levine wasn't writing about couples but about cultures in the thirty-one countries where he studied "the pace of life." But his distinctions about how we organize time, and what might give us offense socially, hold true on the personal level. Of course, people who live by event time don't rely entirely on "spontaneous schedule"; they plan and talk about duration. But if you live in Burundi, Levine explains, and you want to meet in the middle of the day, you might set your appointment time for "when the cows are going to drink in the stream." If you live in Madagascar, and you're remembering how long your appointment lasted, you might say, "the time of a rice-cooking." And if you live in the Andaman jungle in India, and you're remembering when the appointment took place, you might say, "During the scent of the pyinma," as the Andamanese's calendar is constructed around the scents of seasonal trees and flowers. But most importantly, if you're living on event time, you generally allow your appointment or meal or conversation to take whatever time it takes, and if a long appointment makes you late for getting to class or a drink with a friend, no one really minds. But if you cut your appointment short by tapping on your watch, or even pointing towards the cows no longer drinking at the stream, you're likely to be considered rude.

More than a century ago, the railroads brought clock time to the US for good, and we've prospered economically because of it. But for the past century, most in the middle class enjoyed living in event time too. We all understood what Michael Jackson meant in 1979 when he sang, "So tonight gotta leave that nine to five up on the shelf, and just enjoy yourself," and what the Bangles meant in 1986 by "It's just another manic Monday, I wish it was Sunday, because that's my fun day, my I don't have to run day." We had time enough to spare for event time at night, on weekends, and on holidays. And despite the time protestations of Cincinnati, Boston, and Detroit, we had enough range in our tempos for tempos to range across the country (northern locales generally faster than southern, urban generally faster than rural, and cold generally faster than hot) and for tempo differences to be common among people, most conspicuously couples.

But now we're living in a different kind of time. While the new wired time sometimes mimics event time—like when you plunge into a Google deep dive, then look up hours later with strangely haphazard knowledge of the Andamanese scent calendar—and sometimes mimics clock time—with alerts for upcoming meetings you don't want to attend—it isn't really either. Now we've got Lady Gaga singing "Stop callin', Stop callin', I don't wanna think anymore, I got my head and my heart on the dancefloor," because she can't leave that nine to five up on the shelf so easily, and Beyoncé singing, "I shoulda left my phone at home, 'Cause this is a disaster, Callin' like a collector," because trying to escape from wired time is bound to piss off a boyfriend or girlfriend or someone. Even when you're ready to stop running into that empty room to check for a message, you know other people aren't ready to let you stop running. Instead of time that keeps everything from happening at once, wired time gives you the feeling that everything *is* happening at once, and yet that nothing is really happening in the moment because time is splintered into tiny shards, arriving like starlight from different times, none of which is the present moment you're in. Under the guise of efficiency or even fun, often all we're left with are the trappings of efficiency and fun, a kind of perpetual fracturing of the moment, a strange standardized speed. We're back on Taylor's stopwatch in the steel factory, clicking and tapping, marching, or not dancing, a little more to the same drummer, having a harder time slipping into event time at dinner with phones on the table, or into focused clock time at meetings, or back into event time on a slow Sunday afternoon lounging on the couch.

Since Mom got her iPhone, she doesn't ask Dad what time Michael said. She just puts her book down, checks her phone for texts, then checks a few emails, then skims a few articles sent to her inbox, then wonders where her Sunday afternoon has gone.

I REALIZE NOW that when I think of the kind of time I fear losing most, the experiences that come to mind are mostly of being alone. Walks in the woods, or playing sports, or reading. Maybe this is a bad sign.

In my early twenties, during a semester abroad, I dared myself down a glacier in the French Alps. No other skiers visible, thousands of empty vertical feet, a sky newly blue against all that white. I remember gliding over a ledge and feeling my skis carrying hundreds of feet in the air, mind shocked clear by the wind, eyes able to see the smallest things, the fluctuations in the wind-packed snow, the red tram crawling up the side of the mountain, and yet I didn't feel these things separately but all together, smoothly unfolding, my descent down the mountain as though held in place, as though riding on the invisible rails of one long moment. Skiing fast opened a new sense of the mountain, one that forced me to see slowly—to take in more frames per second, to live in speed and slow motion all at once.

With reading, that feeling, that utter absorption, came as a shock. I used to regard reading as a chore, as a strict clock-time undertaking, because it was one: my parents required my brother and me to read half an hour every day. Matt loved the Hardy Boys and would bafflingly wake up before school to follow the latest mystery, the sky-blue bindings of the series forming its own horizon in his closet. As far as I was concerned, real life happened outside, usually in the driveway playing basketball. That's where my dreams of Celtics glory needed only the first resonant bounce of the basketball in the garage to leap into motion, where an imaginary game-clock ran down with pressure on Larry Bird, or Dennis Johnson, or a little-known rookie from the suburbs to hit the winning shot, until Mom was opening the window above the backboard and saying wasn't it getting too dark to see, weren't my hands getting cold, and anyway it was time to come in and set the table for dinner. To get back to the court I served the mandatory half hour of reading with one eye on the clock. Reading for school was no different: work to get done so I could get back outside, and back inside myself, to a kind of time where time seemed to vanish.

So *Dubliners* was an astonishment. The first story we read for Senior English was "Araby," and three paragraphs into "Araby," the narrator describes playing outside:

> The cold air stung us and we played till our bodies glowed. . . . We ran the gauntlet of the rough tribes from the cottages, to the back doors of the dark dripping gardens where odours arose from the ashpits, to the dark odorous stables where a coachman smoothed and combed the horse or shook music from the buckled harness. When we returned to the street light from the kitchen windows had filled the areas. If my uncle was seen turning the corner we hid in the shadow until we had seen him safely housed. Or if Mangan's sister came out on the doorstep to call her brother in to his tea we watched her from our shadow peer up and down the street. We waited to see whether she would remain or go in and, if she remained, we left our shadow and walked up to Mangan's steps resignedly. She was waiting for us, her figure defined by the light from the half-opened door. Her brother always teased her before he obeyed and I stood by the railings looking at her. Her dress swung as she moved her body and the soft rope of her hair tossed from side to side.

My Lord! First the boys playing outside like me, then the "stables where a coachman smoothed and combed the horse or shook music from the buckled harness." The phrase made me want to be quiet, as though certain daily activities might be audible as a kind of music. And then there was Mangan's sister! Where was she in my backyard? And just two paragraphs later, as he's walking through the hostile sounds of the marketplace, through "bargaining women, amid the curses of the labourers, the shrill litanies of shop-boys," he admits that Mangan's sister made him hear a kind of music inside himself: "These noises converged in a single sensation of life for me: I imagined that I bore my chalice safely through a throng of foes. . . . My body was like a harp and her words and gestures were like fingers running upon the wires."

I'd never read anything like it, not even that last phrase, however clichéd it may seem now, and by the time I finished the story, I'd completely forgotten I was in my room, on my gray down comforter, a poster of Larry Bird dunking above me. When Mom called for me to set the table, I remember descending the stairs slowly, feeling unsure of my surroundings as though having returned from a long trip away. Not just from turn-of-the-century Dublin, and seeing Mangan's sister lit by lamplight on her doorstep, but from a different kind of time—in which I wasn't alone, as I'd been in the driveway, but also wasn't quite not alone either. I felt as though I'd found a friend, even though Joyce had lived nearly a century earlier. As though I was now carrying down the stairs, as part of my own chalice, the span of years in between our lives.

But, of course, you don't need to be alone to experience this slower and fuller kind of time. In his memoir *Speak, Memory*, Nabokov describes a walk with his parents through their leafy country estate on his mother's birthday. He is four, his father is thirty-three, his mother is twenty-seven. As he considers these numbers, he is "given a tremendously invigorating shock. As if subjected to a second baptism . . . I felt myself plunged abruptly into a radiant and mobile medium that was none other than the pure element of time. One shared it—just as excited bathers share shining seawater—with creatures that were not oneself but that were joined to one by time's common flow, an environment quite different from the spatial world, which not only man but apes and butterflies can perceive." The adult version of this human awareness of time, one Nabokov conjures again and again in his memoir, is existing in two kinds of time at once—feeling the slowness of event time, of luxuriating in the seawater, while also, through a strange optical effect afforded by that slowness, seeing the whole seaside from above with a kind of survey map of time, aware that the waters are not infinite, and that everyone's time here is limited.

Since returning from the woods, I've felt this dual, almost musical sense of time, sometimes in the least remarkable places. A few years ago, it hit me in a coffee shop near Boston University. It was a groggy Sunday morning, thin winter sunlight suspended above the sugar dispenser. Maybe a cloud shifted outside the large front window, maybe the softly hip music struck some time-altering chord, I don't remember, but for some reason, as I waited for my order, I suddenly saw everyone as beautiful, a young couple with their bedheads and feet touching under the table, two women in puffy black parkas leaning together over mugs, steam catching the sunlight, an old man in a paint-spattered flannel underlining a newspaper—all of us together in time, all of us existing on a morning in no way remarkable other than for the fact that we did exist, that we were here, and that there would come a time when the trolley tracks out the window, and the apartment buildings beyond them, would still be here and we would not.

Maybe the most profound experience of time is to feel this with someone you love. Just a month ago, I was back in Boston and saw my former girlfriend. We hadn't seen each other in the nearly two years since I'd moved to Chicago. We planned to meet in the replica of Thoreau's one-room cabin by the Walden Pond parking lot, mostly as a joke, but also because it was late October and the woodstove would be going. Once I saw her car wasn't there, I didn't want to go into the cabin. I didn't want to lose a moment of seeing her. It was the right decision. When she stepped out of her car, everything came flooding back. The light fierceness of her walk, what her eyes used to do when they found mine.

We walked around Walden Pond, went for lunch, then went for another walk. Nothing had changed: not our feelings for each other, not the difficulty of our circumstances. The only thing that had changed was that we were getting older and we knew it.

We were sitting inside my father's car at this point, parked on a quiet side road beside a meadow, and I felt like I was in high school parked with my girlfriend, and also in my twenties during the summer I lived in a fancy garage nearby, and also in my thirties when the woman beside me had actually been my girlfriend, and also in my forties with the future contracting—the heat going in the car, the Monday afternoon pulling on us from the outside, pulling her to return to work, pulling me to drive back into Boston and catch a flight to Chicago. Once we said goodbye, time would sweep forward, our daily lives would slam into gear. Her shoulder nuzzled against mine, the touch holding us together as we held ourselves back, and the moment felt almost large enough for us to live inside. But time was pulling on us—the limited time we had left together, the limited time each of us had left, apart from the other, to make a life.

THE WAY EXPERTS EXPLAIN a slowed experience of time is through the "pacemaker-accumulator" model. This model suggests that we all have a kind of internal pacemaker: it ticks; it counts the ticks as they accumulate; the ticking can speed up or slow down. At moments of heightened arousal, whether intellectual (reading) or physical (skiing) or emotional (talking in a car with your ex)—the ticking speeds up, as though you're a filmmaker recording more frames per second. You take in more information, but your mind projects these extra frames at the regular speed, so time seems to lengthen and slow down. Imitating this perceptual quirk, the director Arthur Penn shot the famous slow-motion sequence at the end of *Bonnie and Clyde* with four cameras filming at different speeds. When projected at normal speed, the footage shot with more frames per second turned into slow motion: the frames took longer to run than they'd taken to record. Warren Beatty and Faye Dunaway, riddled by bullets, twitched and shuddered tragically slowly—not at the speed of life and death but at the speed of aroused perception and emotion.

It turns out that seeing this way, having more clicks and a slower sense of time in the moment, is deeply satisfying, especially if you're not in mortal danger or saying goodbye to a person you love. The University of Chicago psychologist Mihaly Csikszentmihalyi calls this "the flow state." His research team had colleagues around the world ask people about "optimal experiences"—or when they felt most happy. Everyone more or less had the same answer, not about their favorite activity, which could just as easily be cooking or running or writing poetry, but about its effect on their perception of time. "The flow experience," Csikszentmihalyi writes in his bestseller *Flow*, "was not just a peculiarity of affluent, industrialized elites. It was reported in essentially the same words by old women from Korea, by adults in Thailand and India, by teenagers in Tokyo, by Navajo shepherds, by farmers in the Italian Alps, and by workers on the assembly line in Chicago." The conditions for entering this flow state were the same in every culture: having your body or your mind or both challenged at the far limit of your abilities "in a voluntary effort to accomplish something difficult and worthwhile." You just need to be "so involved in an activity that nothing else seems to matter."

But how do you know if how you're spending your time is changing your sense of time in a way that is ultimately worth your time? What kind of immersion, of feeling that "nothing else seems to matter," builds towards a sense of purpose, towards accomplishing "something worthwhile," and what is just a trip to the land of the lotus-eaters?

THREATS

I recently went online to learn about early seed germination, how plants are springing up in autumn rather than spring. The pattern seemed a possible analogue for how we're post-Ecclesiastes, how we no longer seem to have a time for every season but a single time for all seasons at once—not just with the weather but with our daily lives. I googled, found a few useful articles, and as I was reading, the red-dot alert for an email caught my eye. It was from an old friend, subject header: Tonight! We'd recently had a fight about how I don't respond to emails quickly enough, how it makes her feel ignored, so I clicked. She proposed a movie and included a link to the movie's trailer, so I clicked again. That trailer led me to three more trailers (clever, dull, stunning), three interviews with Lady Gaga (her beauty, her sailor's mouth, her flashes of vulnerability!), two with Bradley Cooper (his restlessness, his charm!), and forty minutes later, I knew nothing about early seed germination and had lost the thread of why I cared in the first place.

Flying down the rabbit hole, I'd felt in a state of flow, getting the extra ticks, time passing effortlessly. But really I'd been tricked into a state of arousal by a digital boy who cried wolf. A recent study by neuropsychologist Sylvie Droit-Volet may help explain why. She showed subjects images of two Degas statues of ballerinas. One ballerina was at rest, hands on hips; the other ballerina balanced on one foot, free leg stretched out perpendicular to the floor, opposite hand extended. After looking at each on a computer screen, the subject was asked to estimate how long the image of the ballerina had lasted. With the ballerina who seemed to be in motion, subjects regularly overestimated, perceiving her interval on the screen as longer than it actually was. With the image of the other statue, they didn't. The subjects' internal clocks were speeding up as they looked at the ballerina who seemed to be in motion, counting up more ticks—so time seemed to lengthen, to slow down. As Alan Burdick notes of the experiment in his book *Why Time Flies*, "In the last few years, Droit-Volet and others have demonstrated that when we embody another person's action or emotion, we embody the temporal distortions that come with it." Our bodies respond to motion with a readiness for motion, and our faces respond to faces with a readiness for emotion. In other words, perception of time is somewhat contagious—our brains and bodies evolved so that empathy starts, for better and for worse, with feeling someone else's sense of time.

So as I'd watched Bradley Cooper and Lady Gaga, my internal clock had adapted to theirs—to the heightened moments of their movie trailer, to the actual motion (forget Degas) and quicksilver emotions—the genuine always threatening to push through the prepared—of their celebrity interviews, and tumbling down, down, down I went. Csikszentmihalyi's definition of flow seems to apply directly: "the state in which people are so involved in an activity that nothing else seems to matter." If I hadn't been trying to understand early seed germination, and if this is the state that people around the world report as the condition of their happiest moments, why limit screen time? Why not tumble down the rabbit hole, why not watch every interview with Lady Gaga and Bradley Cooper, why not just give myself over?

"In the long run," Csikszentmihalyi writes, "optimal experiences add up to a sense of mastery—or perhaps better, a sense of *participation* in determining the content of life—that comes as close to what is usually meant by happiness as anything else we can conceivably imagine." This opportunity to focus your energies, to develop mastery, to experience a sense of participation in a tradition—whether through a sport, an art, a religion, or something else—or to create a kind of tradition of your own through a relationship, doesn't just harness flow moments towards a larger purpose; it's part of what makes those moments flow.

Now, I like Bradley Cooper and Lady Gaga, but the more interviews I watched, the more I knew I was hiding from my own clock, which felt heavy and sluggish because my own work wasn't going well. I was getting hooked on digital time's uppers, riding someone else's clock without the doing the work that was creating its sense of time, so really what I was experiencing was pseudo-flow, at the end of which I looked from my screen with a kind of jet-lagged daze, the sky outside gone dark, the hours never to return.

So much for my early seed germination research. The metaphor suddenly applied to me—the usually ample span of time collapsing, no longer a time for every season. As one of the few articles I read pointed out, "Seedlings resulting from anomalous autumn germination are unlikely to be programmed to withstand a long-lasting alpine winter season." My ideas had become unlikely to bloom.

OF COURSE, many online activities, especially using social media, fall into a deceptive middle ground: we feel as though we're participating, often in important cultural conversation, as we're reading up and having our say. Isn't this a genuine kind of flow, a direct participation in the content of life? Sometimes. If you're a social activist, if you're willing to take what happens online and live it in your life offline, willing to engage in sustained protests and calls to action that actually become actions, then yes. Or if you're gaining strength and support, like my friend whose child has autism, or another friend who was sexually harassed and drew thousands of likes on Facebook for sharing her story as part of the Me Too movement, if you feel heard and respected, then that's invaluable.

But for most of us—and as a writer, I often struggle with this pitfall in my professional life—it's all too easy to mistake calls to action for action itself. If you don't go beyond the first step of posting, then all you're becoming expert in is how to be successful within a conversation held with other people who have roughly the same beliefs, which is more likely to determine the content of a few posts than "the content of life." You might even achieve a kind of moral pseudo-flow, hit the righteousness zone, a feeling of time dissolving in your own high-mindedness. You might thrill to be pointing the smartest, most eloquent, most damning finger. I've certainly been there. But afterwards, when I turn off the screen, I'm rarely left with a feeling of participation or expertise or even community—just a queasy sense of leftover energy, of having been tricked into throwing punches in the dark, my allies and enemies, whoever my allies and enemies really are, nowhere in sight.

WHAT YOU CAN DO

Not long after returning from the woods, I went to an artist residency in rural Wyoming at the foot of the Big Horn Mountains. A cluster of buildings huddled under a stand of cottonwoods. A red one-room schoolhouse had been converted into a dining room; the town's former train depot had been converted into a couple of studios. Writers, artists, and composers came, mostly from New York and LA, to slow down, to watch the deer bending to the long grass out their windows, and to work with the size of that sky behind them. For me the residency felt more like a halfway house—a chance to get socialized again after my two years in the woods, to have some human interaction every night at dinner with eight other human beings.

At the end of our month-long stay, my fellow residents and I wanted to thank the staff for the food and, of course, for the time. As there was no town nearby, and no store to buy a gift, we decided on a collage: the visual artists attached a few small sketches, the writers a couple of lines or sentences, and the lone composer four bars, a kind of abridged score of gratitude. We were like an excited mismatched family, hoping the spirit of our gift would outweigh its obvious lack of artistic merit. In any case, the strangest part of the collage, at least for me, was the music. The composer, a warmhearted, burly man from LA, pointed out the tempo marking above the time signature: *Howie.*

"I'm the tempo?"

"Man," he said, "you are the slowest. You eat slow, you walk slow, you just *are* slow. I'd watch you out my studio window on your walks. It was like fucking Zen meditation. Slower than the deer. It's unbelievable, man. You're slower than *Adagissimo*, slower than *Larghissimo*."

"Those four bars would take a while to play?"

"Like forever. That's how it felt here. Four weeks, but it felt like forever." Then he added, "I was trying to imagine a longer score at that tempo, but you couldn't have a whole score like that. You'd need movements, tempo changes. I don't know how you do it."

Back in Boston, I slowly saw he was right. Just to walk through Harvard Square dizzied me. The curving intersection in front of the Coop had become to my overexposed senses a blur of cars, buses, backpacks, tourists. To step off the sidewalk was a risk. To stand still was a risk; people kept bumping me from behind. I was like a rock in a stream, needing to be dislodged. I needed, as the composer had said, more movements, more tempo changes—over the course of each day, and over the course of my life. My inner clock was still set to the Vermont chickadees, and the morning dew, and the movement of the sun and stars across the sky. There was nothing to speed up for—no reason to eat fast, no reason to walk fast, no reason to try to pay attention to one thing while doing another. My clock had been set by a single slow environment, by how my senses could expand there, and the residency in Wyoming had allowed me, more or less, to stay in that temporality, to stay "really fucking slow." But now I couldn't cross a busy intersection. I couldn't go to a dinner party and eat and listen to the conversation at the same time. I was stuck not in a different time zone but in a different time setting—a rabbit hole I couldn't easily return from, even though I had returned from the woods.

It took months to accept that adjusting to other time settings wouldn't cost me what I'd learned in solitude, that being Zen, as the burly composer put it, probably meant being the water rather than the rock, and it took years to regain the top end of my time-settings range—to be able to go fast comfortably.

Life at many speeds. With clock time, event time, and even digital time now and then. Life at the right speed, in the right temporality, for whatever you're doing.

David Foster Wallace, doing an interview for a *New York Times* profile, once said to Roger Federer that watching tennis on TV doesn't do justice to the speed of the game. Federer had a strange response: "For the first time since I have come here to Wimbledon, I went to see a match on Centre Court, and I was also surprised, actually, how fast, you know, the serve is and how fast you have to react to be able to get the ball back. . . . But then once you're on the court yourself, it's totally different, you know, because all you see is the ball, really, and you don't see the speed."

It's something I still struggle with—being able to slow my attention to listen to a friend on the phone, or to speed it up to skim through the supermarket no matter the Elton John song playing, or to slow it down to teach. And trying to orchestrate that shifting sense of time so it can come into harmony, can become a kind of dimensionality, with the unknown amount of time I have left here.

THE SUMMER I WAS NINE YEARS OLD my family took one of the great American vacations. After visiting Yosemite, where I saw my first deer in a meadow of high grass and more stars than I'd ever imagined existed, and after touring Los Angeles, where I crouched to touch the stars on Hollywood Boulevard, my dad turned the rental car towards Las Vegas. After the thick heat of the desert, our hotel felt like a dream. Circus Circus had powerful air conditioning, wall-to-wall carpeting, carnival games for kids, and so many screens flashing and beeping I felt like I'd tumbled inside a video game. That next morning my mom was jubilant and exhausted—her hot streak on video poker had carried until one in the morning. That afternoon my brother and I won stuffed animals at a basketball game. By dinner uneasiness was setting in. The constant thrill potential was turning us a little grouchy, like we had to keep catching the thrill but couldn't. I asked my dad why there were no windows. He said so people wouldn't know what time it was. That way you can just keep playing, Mom said. Dad gave her a look.

Two mornings later, when we made it back outside, the daylight was overwhelming. The heat grabbed for my ankles through my socks. I felt slightly transparent, like when you step outside after a long illness. But the shimmering horizon was good to see—we were back under a sky, back with the possibility of change. Everything wasn't happening at once anymore, and each little thing seemed to have all the time it would need.

CHAPTER 3

Frames

ENDANGERED TRAITS

Firsthand experience (First from Middle English, *fyrst*, akin to Old English, *faran*, to go; Hand from Old English, *hand*; and Experience from Latin, *experiri*, to try)

Truth compass (Truth from Old English, *treowth*, fidelity; and Compass from Old French, *compasser*, to measure)

Wonder (from Old English, *wundor*, akin to Old High German, *wuntar*, wonder)

BACKGROUND

Just as I snapped on the car radio, a man said, "Our challenge is to wake up each day and say, 'How can I experience the world better?'" It was a misty night, headlights luminous tatters in the trees, the road abandoned—the kind of drifting night you might receive intelligence from afar. The man's inflection was odd, no clipped urgency of a news reporter, no studied vulnerability of a live storyteller, its defining characteristic hypnotizing earnestness, somewhere in the gray continuum between camp counselor and cult leader. I leaned towards the glowing dash, turned up the volume. The voice continued, "Picasso once said, 'Every child is an artist. The problem is when he or she grows up, how to remain an artist.' We all saw the world more clearly when we saw it for the first time, before a lifetime of habits got in the way." After rapturous applause, the announcer came on and said we'd been listening to a TED talk, delivered by Tony Fadell, the man who'd helped design the iPod and the iPhone.

The misty trees grew more disorienting, more parallel world-ish. Maybe I'd missed something. Before I'd tuned in, maybe Fadell had defined "experiencing the world better," maybe he'd coaxed out of hiding the question of values inside that phrase. Maybe he'd contextualized the notion that "we all saw the world more clearly when we saw it for the first time"—by explaining that it takes infants about six months to recognize their parents by sight, and often another several decades to see beyond their parents' framework of the world. Maybe he'd even said that seeing clearly is an ongoing balancing act between seeing within a framework and seeing beyond a framework, between seeing with knowledge and seeing with wonder. And he must have shown a delicious sense of irony at being a designer of the iPhone, one of the mightiest habit-forming devices of our age, talking about habits getting in the way of seeing the world more clearly.

I sped through the overhanging trees riding a mild TED mysticism, as though on a mission into another dimension, eager to get home and google and understand. But when I watched the full talk online, it turned out I'd missed only an awkward opening, in which Fadell describes a scene from *The Blues Brothers* with painful comic timing, then talks about the irritating sticker on pieces of fruit and how we become habituated to such irritations. The talk's main idea was that the way to "experience life better" was to see the world more clearly by becoming aware of these sticker-on-fruit irritations and working to limit them. The only other things I'd missed were the popularity of the talk (over two million views), and the talk's title: "The First Secret of Design Is . . . Noticing."

SNOWY DAY, EARLY APRIL. The hemlocks edging the park stood arrayed in snow and rime, like old men given an unexpected opportunity to show off their dignity. Along the narrow walkway, a receiving line of maple saplings looked dumbstruck, their mouse-ear-sized leaves bedazzled with ice crystals.

It was mid-morning; I was alone on my morning walk, the same morning walk I've taken every morning since returning from the woods. The daily routine helps me slow down, become more relaxed and observant before sitting down to write—helps me notice the weather, the birds, the other people going about their day, so once I'm back at my desk, wandering around my mind, I can notice the weather, the birds, and the other people there too, and wait for the quiet voice, the one with questions, back by the trees.

But the morning constitutional, and its invisible way of helping me experience the world better, has gone out of fashion. One morning a few years ago in Newburgh, two speeding police cars crested a quiet road near my aunt and uncle's house, and the officers jumped out to ask what I was doing: a woman had noticed "a strange man, not really going anywhere, just looking around." One morning two years ago in Boston, a woman smiled at me over her earbuds, "For an exercise walk, you're really going to have to walk much, much faster." And one recent morning in Chicago, an old man said helpfully, "You should get a dog, that way you might feel like you're doing something useful."

When I glanced up now from the glittering saplings, I noticed a minivan idling on the side of the park. The tinted window slid down. My neighbor John was grinning at me, his breath smoking in the cold. We don't know each other well, but he lives near the park and likes to talk.

"Having a flashback to the woods?"

The air between us was cool and fresh, a kind of champagne. For a moment, I imagined being back in the meadow by the house in Vermont, the smell of woodsmoke drifting from the chimney, and what it would mean to see him there.

"That would make you a crazy man in the woods too," I said.

His minivan idled, exhaust pluming. He blinked once, smiled at the idea.

I WONDER WHAT THOREAU would say at his TED talk about seeing the world more clearly. After his Harvard graduation, he became curator of the Concord Lyceum, a public forum for lectures on topics ranging from botany to education, and he himself later became a lyceum speaker. It's not hard to imagine him with a clear TED microphone fitted over his neck beard, the massive screens behind him saying: TED Dead, Thinkers of Yesterday, Challenges of Today.

"Ladies and Gentlemen, when I first received the invitation to speak, I must confess, pleased as I was, some of the curator's letter eluded my understanding—for instance, the polite imperative: 'Please drill down to deliverables.' Am I to understand you view my ideas as buried underground, like a vein of gold or healing spring? Nevertheless, I understand the thrust of his letter, and I must agree: the truest critics are not critics but innovators. As I wrote in *Walden*, 'I know of no more encouraging fact than the unquestionable ability of man to elevate his life by a conscious endeavor. It is something to be able to paint a particular picture, or to carve a statue, and so to make a few objects beautiful; but it is far more glorious to carve and paint the very atmosphere and medium through which we look.' In this spirit, I shall gladly propose some 'apps.'

"My first proposal is LifeSuck, a daily planner designed according to the premise 'I do not so much wish to know how to economize time as how to spend it . . . How to live. How to get the most life.' Employing an algorithm, as the mass of men seem so fond of algorithms these days, that calculates 'the cost of a thing' as 'the amount of what I will call life which is required to be exchanged for it, immediately or in the long run,' LifeSuck will estimate how much life an activity will suck out of you, versus how much life will it allow you to suck. It may even counsel you to stop using apps altogether, but that is not my affair."

[Anxious TED talk laughter.]

"My second proposal concerns daily travel, which ought to be—as my rambles always were—daily communion with the world. Therefore, my update to Uber will be Ubermensch. As Mr. R. W. Emerson has explained to me, Mr. Nietzsche's 'ubermensch' is a reworking of Mr. Emerson's own "Over-Soul," a conception of the deep spiritual connectedness of all things. Accordingly, Ubermensch will take into account your physical and spiritual destination, your internal and external weathers, and predict whether you're more likely to find communion and inspiration walking a few extra streets with your friends, riding the public bus, or simply ordering an Uber."

[Uneasy laughter. *Was that funny?*]

"To assist with your non-physical orientation, you may employ my update to Shazam, which will identify not the music playing around you but the music playing inside of you. If a man does not keep pace with his companions, perhaps it's because he hears a different drummer. Let him step to the music he hears, however measured or far away. With Drummerzam, you will hear your different drummer on your own earbuds, as you call them. Indeed, perhaps your brain will blossom!"

[Hearty applause.]

"Use of these applications, day after day, year after year, will undoubtedly change the very medium through which you look, thereby improving the quality of what you see, hear, and think."

I READ A REVIEW in the *New York Review of Books* one recent morning of the Fischli and Weiss exhibition at the Guggenheim, in which the reviewer called them "wry magicians—and magicians with an underlying moral bent." What they're up to with their little clay sculptures, their refrigerated snowman, their toys and accordions, and with their film, *The Way Things Go*, which is thirty minutes of pure Rube Goldbergesque action with daily objects as the stars—what they're up to, writes the reviewer, is "a kind of reclamation of the ignored."

Then I went for my morning walk. A gray windy day, the rain spitting. By the bakery, a plastic coffee-cup lid skidded up onto its edge and began wheeling perilously across the street. Its path doomed, a car approaching. But the wind kept blowing, like a children's book drawing of the wind, and the lid kept going, jerking itself this way and that, and it zipped all the way across the intersection, careening into the far curb with a jaunty hop, as though letting out a victorious yip.

It seemed straight out of a Fischli-Weiss exhibit. Without thinking about the article, I'd been seeing through its framework, doing what it had suggested—reclaiming the ignored. I wonder what I would have seen, or not seen, if I'd read a different article over my cereal.

My high school girlfriend once gave me a Valentine's card: "You're the last person I want to talk to before I go to bed, and the first person I want to talk to in the morning." I felt the same way about her. We wanted our relationship—our ongoing conversation and references and inside jokes—to be the frame for thinking about what had happened that day, and the frame for what we'd be attuned to the next.

It was a way of sharing a common window, even when she moved to Montreal, even when we lived hundreds of miles apart.

THIS MORNING MY WELCOME SCREEN, courtesy of Windows, offered me its own meditation: "Create. Edit. Do. Impress." To the screen's credit, it was bit insecure about the quality of the suggestion. In the upper-right corner, it asked, "Do you like what you see?"

Every day it slips me another note. Every day a new welcome photo of a scenic mountain or lake—which makes me, especially if I'm having a bad day, want to think of my screen as she, just like that sad, lonely guy in the movie *Her.* Is she hinting at a vacation? Is she trying to inspire me to work harder, a kind of digital muse? Some days there are just little reminders about ways to make our relationship function more smoothly; other days there are reminders of the value of her speed and efficiency, which themselves are reminders of how much I need her.

Microsoft is taking Thoreau's advice for me—trying to help me see the world better.

What's uncanny is that Microsoft, Apple, and Google all sell their products on this promise. If a recording existed of Thoreau saying, "It is something to be able to paint a particular picture, or to carve a statue, and so to make a few objects beautiful; but it is far more glorious to carve and paint the very atmosphere and medium through which we look," you can bet one of the companies would have paid Thoreau's estate a fortune to use it as a voiceover. Apple's ads have always sold a vision not of how you will be seen—as ads for nearly every other product do—but by selling a vision of how you will see. From the early "think different" ads, with their austere black-and-white photographs of iconoclastic geniuses like Einstein and Picasso, to the point-of-view ads showing hands holding an iPad with a swirl of stars vivid on the screen, what Apple promises is the ability to see more richly.

Microsoft, catching on a bit late, now has an ad that ends with the tagline "Give Wonder."

And Google has a product called Wonder—which enables you to "use natural language to recall things you're wondering." So you can say: "Wonder, what was George Orwell's real name?" But however *1984* it is that Google can collect a history of what you've been wondering, which it can then sell to advertisers, the product is really just a spoken version of how we already use Google and how they already use us.

Just two nights ago, on the phone with my friend Andrew, the movie *St. Elmo's Fire* came up. Really what happened is I started to sing. Andrew and I were college roommates, we're still best friends, and like most people who fall into both categories we begin our conversations with any number of peculiar rituals, sometimes trading names we just happen to like—Chaka Khan,

Scheherazade, Jaime Yzaga—and sometimes singing snippets of old pop songs—"Running on Empty," "Free Falling"—as a shorthand for how the day or week is going. The other night he said he'd been in constant motion, so I launched into "Gonna be your man in motion, All I need is—," even though those were the only words I knew until "St. Elmo's fire." I have a terrible voice, but Andrew caught the gist. "Who was in that movie again?" he said.

"Emilio Estevez. Demi Moore. The really good-looking guy who was in *West Wing*."

"Rob Lowe?"

"Right, he was in it. Do you remember that scene when Emilio Estevez drives up to the mountains to see Andie MacDowell?"

"Sort of." I thought maybe Andrew was getting some applesauce out of the fridge. Then his voice came back stronger but more distant—he'd put me on speaker. "Emilio Estevez, Rob Lowe, Andrew McCarthy, Demi Moore, Judd Nelson, Ally Sheedy, and Mare Winningham. Who's that?"

"The one whose mother whispers *cancer* at the dinner table."

He didn't laugh, his attention was gone again, and then he listed the cast of *The Breakfast Club*. If someone was making a movie of us, we'd have been split-screen, Andrew looking at his phone while I was looking at my ceiling, trying to remember *St. Elmo's Fire*. I was enjoying letting one image lead to another—Andie MacDowell's cabin in the woods, the bar where the friends hung out, the weird scene with Demi Moore in her empty apartment—and I was curious to hear what moments had stayed with Andrew, and even to hear him speculate on why. But since he jumped straight to the facts of the movie, there was no chance for us to talk about our memories of it. He'd been in constant motion, as he said, needing to get answers as quickly as possible all week, so it's no surprise that was the mode he was still in. But wonder Google-style bypasses personal experience, bypasses how our memories are imperfect in personal and even wondrous ways. I didn't really care about the full cast list. I cared about what each of us did or didn't remember, and how browsing those memories could be like flipping through a photo album of our high school minds—a way to think about what we thought, to try to see what we saw.

What's even more disturbing is the way Google seems to want a monopoly on the whole wonder market. They couldn't buy Thoreau's voice, but they did manage to raise Mister Rogers from the dead. The ad for the Google Pixel 3 opens with a freckle-faced girl gazing up at a gorgeous blue sky and Mister Rogers crooning, "Did you know? Did you know? Did you know that it's alright to wonder?" By the time he hits the word "wonder," the image has moved from the sky to a boy looking at a praying mantis through his phone, with an entry that says "Mantis." Lovely images continue in the same golden light—hands picking up sand at the beach, an old man peering at the sky through a window, a flock of birds on the wing—and each time Mister Rogers hits the word "wonder," we see the phone in another person's hand, until the chorus comes around—"Did you know? Did you know?"—and we see the phone on "know," then again on "know," and then finally on "wonderful." The substitution is complete. The image of wonder has gone from a girl being awed by nature, to a boy acquiring information about nature, to people simply holding the phone itself.

Contrast this version of wonder with how the speaker in Mary Oliver's poem "The Summer Day" considers a grasshopper:

This grasshopper, I mean—
the one who has flung herself out of the grass,
the one who is eating sugar out of my hand,
who is moving her jaws back and forth instead of up and down—
who is gazing around with her enormous and complicated eyes.
Now she lifts her pale forearms and thoroughly washes her face.
Now she snaps her wings open, and floats away.
I don't know exactly what a prayer is.
I do know how to pay attention, how to fall down

into the grass, how to kneel down in the grass,
how to be idle and blessed, how to stroll through the fields,
which is what I have been doing all day.
Tell me, what else should I have done?
Doesn't everything die at last, and too soon?

Real wonder tends to come, even for adults, when the unknowable flashes through the known—the mystery of our place in the universe made tangible in a meteor shower, or a sudden silence in the trees that seems to let us in, or even in a grasshopper. But Google makes everything appear fully knowable, indeed already fully known, so all we need to do is reference and cross-reference intelligently and efficiently.

Wonder is in your hands! Wonder isn't a potent swirl of curiosity and awe. Wonder doesn't take time, or have anything to do with the unknowable, or grace you with the humility to kneel down in the grass. Wonder is control. Wonder is over twenty-four million results for praying mantis in under .5 seconds. Wonder is the answer, not the question, and not the questioning.

Look with me through this window, boys and girls. This is how we see.

WHY IT MATTERS

The first time I thought about "experiencing the world better" was at a Picasso exhibit at the Museum of Fine Arts. There was a booth outside the exhibit where you could rent headphones, and no one seemed to be entering the exhibit without them. The woman behind the desk wore headphones up both arms like bracelets, looking like a scarecrow for robot birds.

"Are the headphones part of the exhibit?"

"They're Acoustiguides. They tell you what you're seeing."

I'd been in a play in college about Georgia O'Keeffe, and remembered her character, played by a friend of mine, saying, "Nobody sees a flower really—it is so small—we haven't time, and to see takes time, like to have a friend takes time." I always liked when she said that line, thought of our friendship, the nights going to get pizza at Tommy's. Did I want a voice in my ear telling me what I was seeing while I was still taking time to see? Would it deepen the experience to hear an expert? And what did deepening the experience really mean? Was I hoping to learn more about Picasso's Blue Period, or to be moved to empathy, or to be reminded of something in my own life? Or was I hoping for a kind of bonus ball, something I didn't even know to look for? How did you find an answer to a question you didn't know how to ask?

The man behind me pushed forward. I decided to skip the headphones. Turning the corner and entering the climate-controlled air of the exhibit felt comfortable without anything on my ears, not because I knew what I wanted to experience but because I knew how I wanted to experience it—directly. The whole set-up of the exhibit was itself a frame, yes, but I tried to go at my own pace, ignoring the Acoustiguide migrations. I tried to take time with the paintings that intrigued me, the way I would with a friend. I came to *Woman Ironing*. I looked at her gaunt shoulder and the burdened yet tender angle of her head, at the length of her fingers, at how her whole careworn body seemed collapsed upon the iron, as though it were a plinth holding her up, or pulling her down, or somehow doing both. I looked at the colors, the indeterminacy of the blues and grays. I thought of an ex-girlfriend who was an artist and the way her shoulder looked when she worked at our kitchen table, its jutting intensity, the way it would reveal more about her suffering during those hard years than her face. Finally, I read the placard on the wall, then looked again, and appreciated more of the blue in the woman's thin dress, more of the gray in her hair.

A week later Andrew went to the same exhibit, then gave me a call. "The Acoustiguide was great, right? So helpful."

"I didn't get it."

"Big mistake," he said. "Huge!" He began to describe how Picasso came to paint *Woman Ironing*, how he'd suffered as a poor young artist in Paris, how he'd identified with the working class but hadn't fallen into the trap of glorifying the working class as other painters had. The painting wasn't just a concentrate of empathy; it was a groundbreaking concentrate of empathy. "You see so much more once you get the history," Andrew said.

I felt foolish, outmoded. My frame, or general lack of a frame, had helped me to see things but prevented me from seeing others.

Maybe I had missed out.

So I went back to the museum, bought a book on the exhibit and another called *Picasso on Art*, which is a collection of his quotes. I went home, read for a couple of weeks, then went back. I noticed different things than I'd noticed the first time, but what struck me most was how *Woman Ironing* came alive again only as I stood in front of it: the space between the deep shadow of her collarbone and the woodstove huddled behind her, the gradations of the blues and grays, the space between the painting itself and where I felt comfortable standing to look at it. What I felt was the presence of the painting, really, a presence that couldn't travel fully in memory. My new frame wasn't much more informed than my initial one, but I realized that no matter how much I learned about Picasso, there would be no end to looking. Whatever frame I came with, the painting, just like a person, would never be contained.

About this mystery, the mystery of what he calls a painting's holiness, Picasso said, "You can search for a thousand years and you will find nothing. Everything can be explained scientifically today. Except that. You can go to the moon or walk under the sea, or anything else you like, but painting remains painting because it eludes such investigation. It remains there like a question. And it alone gives the answer. It has that good fortune. And that misfortune. And we too."

The pale grayish blue of the woman's dress now seemed a reminder of the blue of distance, the way even mountains fade to blue from far away, and the deep charcoal blue of her eyes seemed a suggestion of her having been overwhelmed by distance, of having been banished, even in a Parisian laundry, to being able to see and be seen only from far away.

You could know her, but only so much—and knowing that was one of the keys to knowing her.

AT A CROWDED PARTY LAST FALL, a friend who works for a top technology firm told me he was working on A.R. Steve's shorter than I am, and we were pressed close together, our glasses of wine terraced between us, so he could raise his arm to drink only when I drank. The party was echo-chamber loud. I thought he said he was working on air, and I wasn't sure if he was joking.

"Didn't know it could be improved."

"What?"

"The air."

"No. A.R.!"

"What's that?"

"Augmented reality. Like Google Glass, but better. It'll probably be contact lenses. You just wear them all the time. You get information about whatever you're looking at, plus updates for your social media, your texts, everything."

Steve has a straight man's delivery—he sometimes writes comedy—but he wasn't joking. "So what would my A.R. tell me now?"

"Anything. That this wine got 237 mediocre reviews. That I'm a stud on Twitter. Or the spread on tomorrow's Pats game, if we were being bros and talking about the Pats."

"It would hear what we're saying?"

"Pretty cool, right?"

"This is going to happen?"

"It's next level. You won't have to look down at your phone."

I looked down at my drink.

"There will be some issues. Like how much advertising will be allowed, and what it would mean if someone didn't wear the lenses." Steve nudged me. "Like what would happen to them."

"What would happen to them?" I pictured an A.R. resister in the stocks, information being flashed before his eyes. Even my public-shaming references were out of date.

"I'm just messing with you. Sort of. It won't be easy to live if everyone else has A.R. and you don't. It'd be like everyone was on information steroids, and you'd be the skinny kid in the corner, unable to compete." He said this matter-of-factly.

Eventually we unterraced our arms and I made my way to the balcony for some air. It was a warm night, mist floating under the streetlight's glow. The party hummed behind me. A black cat skulked along a low wall. What Steve was describing wasn't just a metaphor, like a lens for the inner eye, but a real contact lens, a visual Acoustiguide overlay, one designed not by knowledgeable life docents—who would they be, anyway?—but by mostly young men, likely in skinny jeans, trying to make their first billion.

I imagined the station wagon in the driveway across the street labeled with make, model, and resale value, the woman moving behind the curtain trailed by a floating CV. I remembered the old woman Mrs. Dalloway watches in the apartment building across the street from her own, a woman climbing her stairs, then parting her curtains and looking out her window not knowing she is being seen—how Mrs. Dalloway feels something solemn in the woman's privacy, especially as the old woman retreats into the back of her bedroom, how Mrs. Dalloway sees her as a kind of mirror for what is unseen and unknowable in herself. "And the supreme mystery . . . was simply this: here was one room; there another." Distance, and the awareness of personal distance, of the old woman's ultimate unknowability, was necessary for Mrs. Dalloway to feel close to her and also more at ease with her own intermittent bursts of loneliness. How are we going to feel that sense of communion, that common isolation of having a consciousness separate from everyone else, if our eyes are being flooded with the impression that everyone and everything can be known?

I imagined seeing the precise speed of the wind and the weather forecast, an advertisement for the company behind the flagstone walkway, a flag noting an open McDonald's 1.2 miles away. Even the smoky streaks of cloud above the roof couldn't escape labeling, stratocumulus, as though stuck on a collector's pin. I felt like a King Midas of the Eye, everywhere I looked hardened into information that might have been useful, if everything around it hadn't hardened too.

A FEW WEEKS AGO, after five days and nights of April rain, my morning walk took me along the community golf course, which had effectively been turned into a nature preserve. A great blue heron coasted over my head, long legs trailing behind it like an airplane banner advertising the morning. Robins road-runnered forward on the wet grass, stumbling to a halt when I glanced over, as though playing red light, green light. The grass of the seventeenth fairway had become a one-tree island, the oak's trunk like a barbershop pole, golden shimmer off the makeshift pond climbing it steadily. Not far from the tree, an odd bird was swimming around the periphery of the water.

A bright red-and-white beak, a black-and-white face, and a black crest with a smart white stripe, giving its head the shape of a Trojan army helmet. I'd never seen anything like it. It paddled a little, got nervous, waddled up on land, then flew up into a tree.

Back at my apartment, I struggled to locate my *Peterson Field Guide*, a book I almost never consult. Then there was the challenge of finding the bird within the book. But there it was, on page 68, the wood duck!

> Bizarre face pattern, swept-back crest, and rainbow iridescence are unique. Fairly common.

Bizarre, absolutely! Fairly common, who cared? The wood duck was no great discovery, but observing it, wondering about it, and then learning about it had been strangely thrilling.

The golf course soon dried out, and the wood duck vanished. I kept walking, hoping to chance upon another exciting creature. Day after day, just squirrels, cardinals, then the arrival of red-winged blackbirds. But just last week, one very bright morning, a golfer stood beside the edge of trees on the seventeenth tee. Over his shoulder, he said to his buddy, "Quiet, you'll scare it away." Quietly, I drew closer.

The man turned, zipped up his fly, and walked jauntily back onto the course.

Even so, I felt a sense of wonder—in my head, there was a voice like a little boy's: that man was pissing in the trees!

THREATS

T. J. Clark makes perhaps the most poignant case for the significance of frames. An influential art critic, he found himself during a six-month fellowship at the Getty Museum doing something strange. He'd planned to work on Picasso during the wars (the notes for the project were in his car), but instead of looking at Picasso, Clark returned every day to the same two paintings by Poussin, *Landscape with a Calm* and *Landscape with a Man Killed by a Snake*. He wasn't a Poussin expert, he had no project in mind, he was "just looking." Art critics don't just look. This departure worried him, this non-project that could have been viewed professionally, and personally, as a kind of surrender, and not in the positive sense.

But Clark turned it into a project. He kept a journal recording what he noticed day after day, as the light in the gallery shifted with the LA weather, as his own thoughts and questions shifted, as what he saw in the paintings shifted. Perhaps at first he didn't know what he was looking for with his new way of looking. At first it seems he's intrigued by looking without an informational frame. But Clark is hardly making the case against knowledge or informed looking—his looking gives rise to questions, for instance about Poussin's relationship to his patron, which he will need to look up. But the questions arise from the looking, rather than the looking arising from preconceptions. Which gets to what's really driving Clark: the attempt to look at the paintings on their own terms, in their own visual language, and to document how looking slowly again and again helps him appreciate the ineffable in the painting—the play of light and dark, the poses of the figures, the spatial relationships between them—all of which he paraphrases in his journal, knowing they cannot be paraphrased, knowing their ability to communicate exists in a realm beyond language. They must be experienced firsthand.

While Clark's work may sound like an extended swoon into two landscapes, like it has nothing to do with the complexities of the city and country beyond the gallery walls, his project was deeply political. A growing percentage of the daily images most people saw, even by 2000, were meant to be absorbed in a glance. You weren't supposed to consider the relationship between a Coke and a Smile, or between Gatorade and your chances of being like Michael Jordan. They were just supposed to be instantly familiar, so you could reach for a Coke or a Gatorade as though remembering a pleasant personal memory. You sped past an image on a billboard, glimpsed it on TV, flipped past it in a magazine. To look closely wasn't the way the language worked, wasn't the way your eyes were supposed to work. Just the opposite—to slow down and pay close attention would turn fluent speakers, the magazine and billboard and television images, into stuttering hucksters. A sugary drink, corrosive to your teeth, was supposed to make you want to smile? Another sugary drink was supposed to make you play and *be* like Michael Jordan, the best basketball player in the world? Seeing the daily world through Clark's slow visual language, you would have felt like you were in a foreign country, surrounded by communications that made no sense.

In a journal entry from late February 2000, the anger swirling beneath Clark's project breaks through his daily meditations:

> I have not been thinking in general terms about the turn in my writing this whole project represents: the writing itself has hardly left room for that. But I feel no unease about the turn, if it is one—above all no uneasiness about its political point, its political responsibility. I did not start this book from the memory of Y2K for nothing. We are living, I reckon, through a terrible moment in the politics of imaging, envisioning, visualizing; and the more a regime of visual flow, displacement, disembodiment, endless available revisability of the image, endless ostensible transparency and multi-dimensionality and sewing together of everything in nets and webs—the more this pseudo-utopia presents itself as the very form of self-knowledge, self-production, self-control—the more necessary it becomes to recapture what imaging can be: to suggest what is involved in truly getting to know something by making a picture of it: to state the grounds for believing that some depictions are worth returning to, and that this returning (this focusing, this staying still, this allowing oneself to respond to the *picture*'s stillness—everything hidden and travestied, in short, by the current word "gaze") is a form of politics in itself, meeting other forms head on. Only professional twenty-first-century intellectuals, chained to their image-displacement machines like lab animals to dispensers of morphine or 220 volts, could be so blind as to think that looking at Poussin is a matter of nostalgia, or elitism, or some such canting parrot-cry.

Read the passage aloud, and you can feel Clark's anger swirling in a rhythmic updraft, gaining momentum, until it reaches escape velocity and bursts from the quiet gallery out into the world beyond. It would be easy to dismiss as the raving of a guy spending too much time alone if his warning about a regime of visual flow, the danger of its "ostensible transparency" and the illusion that it could lead to knowledge, let alone self-knowledge, didn't so eerily foreshadow the hacking of the 2016 presidential election and the era of the selfie. He's defending a visual language—a way of making, looking at, and remembering images—that takes time, that requires context, that rewards focus and analysis and return. A visual language that is irreducible, just as Picasso said, just as O'Keeffe said. He's defending it as a mother tongue, not just for art critics but for all of us, as a way to get to know the world and each other more truly.

I felt a similar anger when I came back from the Vermont woods. I remember watching TV alone in a friend's house in New Hampshire for five hours straight, entranced by how everything on the channels corresponded with each other, how Jordan went from the sports highlights on the news to the advertisements, how the speed of cuts and camera angles corresponded across the channels, how what I was looking at seemed to be a unified world, with a unified assumption not just of what to look at but of how to look. The sitcoms, the sports, the movies. That world so enchanting and disorienting I couldn't stop until my head was throbbing. That evening I took two Tylenol and draped the TV with a musty beach towel. I didn't want to be tempted to look through the TV's window. I didn't want to be looking through its way of looking.

OVER TIME MY FULL RANGE of visual languages returned, and television became an occasional pleasure rather than a threat. I really had missed movies and watching the Celtics. But now as more and more of my daily life comes through my computer—the news, Celtics games, movies, emails from colleagues, students, friends—I struggle to maintain a visual diet that feels healthy. Just the other day I was reading an essay on *Slate* by a friend of mine. She'd spent time with an old woman who'd lived around the corner from Anne Frank as a little girl in Amsterdam; the woman told my friend, and has said publicly, that Anne was "a brat." The main idea of the piece was that we shouldn't need to martyr the victims of the Holocaust, that Anne may well have been smart and brave and also kind of a brat. But as I was reading down the screen, picturing Anne in the attic, adjusting my version of her to allow for a little more of her humanity, my eye came upon a photograph of a sultry woman with tousled hair and sleepy eyes, lounging by a window in her underwear. A few more paragraphs down, there she was again, still in her underwear, the glimpse-language of the advertisement all too clear. My mind felt harried along, as though the cobbled road I'd been traveling down, a road my friend had carefully built through her writing, was really a highway, and cars were backing up behind me, honking at me to travel faster. My visual imagination, which needed time to picture Anne actually acting her age, didn't stand a chance.

Even in art museums, the visual language Clark was trying to protect is endangered. On a recent outing to the Art Institute in Chicago, I noticed a beaming young couple, probably honeymooners, snapping selfies in front of van Gogh's self-portrait from 1887. He painted it three years before his death in Auvers-sur-Oise. His eyes are leaden, pained with their own intensity, the pink at the inner corner raw. My friend nudged me. "Think he was smiling before they came?"

The couple checked their image, posed, smiled again.

Something similar was happening in front of nearly every Monet, in front of several Seurats. I thought of my cousin who last Thanksgiving predicted that one hundred years from now everyone in the world would speak only English. He said we'd be better off that way.

A single visual language seemed already to be arriving. Not one that teaches us to look more deeply, but one that teaches us to glimpse.

———

AFTER MY EYE ACCIDENT I thought a lot about blind spots. I'd sit on my bed in Adams House watching friends walk up to class. My brain hadn't registered the change in my peripheral vision, and I'd crashed too often into some unsuspecting person on my right to continue using the sidewalk during class rush hour. When I did make my way to lecture, after the sidewalk spread out again, I'd sit in the back row all the way on the right to have the full room in front of me. That way there was nothing I couldn't see.

I was twenty years old, trying to figure out what to do with my life, and what began to frighten me was other kinds of blind spots. Seeing happened in the brain, my ophthalmologist told me, not in the eye, and I began to have odd daydreams during my appointments. What if Dr. Grosskreutz put up an eye chart for the inner eye—how would I fare on a test for insight? Or a test for seeing the big picture? *Tell me how these sideways letters relate to the cosmos*, she might say. The thought of being prescribed corrective lenses frightened me as much as the thought of being allowed to walk back outside with whatever blind spots I had. I imagined the old woman sitting next to me looking up from her magazine. "Walter's here for the racism screening. He insists he's fine, but I do think he could use a prescription."

What worried me, really, was missing some truth about life I'd always been missing. What worried me was being blindsided and getting hurt again.

WITHOUT CONNECTING THE DOTS about digital life, Daniel Kahneman explains several of our cognitive blind spots in his book *Thinking, Fast and Slow.* Our forty-thousand-year-old brains were built for efficiency, for making quick, intuitive decisions about avoiding danger and finding shelter and food—which means they were built to take shortcuts. Our brains evolved to keep us safe, not to give us a totally accurate picture of the world. Maybe I'm overly aware of the dynamic between vision and danger, but I'm hardly the only one aware that if you play on those blind spots with a visual language ideally suited for manipulating them, we're in serious trouble. The cognitive shortcuts that evolved to keep us safe begin to make us unsafe.

First there's the availability bias—if you've seen a few stories about tornadoes, you're naturally going to start fearing tornadoes and overestimating the probability of a tornado spiraling your way. Then there's the affect heuristic—if those stories featured devastating images of homes ripped open like dollhouses, you'll think the probability of an impending tornado is greater still. Kahneman explains, "The world in our heads is not a precise replica of reality; our expectations about the frequency of events are distorted by the prevalence and emotional intensity of the messages to which we are exposed."

Now more than 60 percent of Americans get their news from social media. Everyone from the *New York Times* to NPR is funneling money from long-form journalism to more video content, aided in part by a $50 million grant from Facebook. The images flicker past on our feeds. The smiling daughter of a friend on her first day of school, a hellish forest fire in northern California, a car insurance advertisement, Trump making a windblown speech beside Air Force One—each image less real, less distinctive, all a part of the same simplified narrative, but about what?

Hypnotized by this visual stream, we fall back on shortcuts, on stereotypes, on sorting information by our fears, by the child's question of who are they and who are we, rather than by the adult questions of what is the situation and what can we do about it. Algorithms keep nudging us towards more captivating, more extreme versions of our political leanings—show interest in conservative content and autoplay on YouTube will lead you from videos of Donald Trump speeches to white supremacist tirades to Holocaust denials; show interest in liberal content and autoplay will lead you from videos of Hillary Clinton and Bernie Sanders speeches to extreme left conspiracies about the US government planning the September 11 attacks. We lose a common sense of the facts, and we can't use reason to debate issues based on fact, which leaves us to trust only our own clans.

Now throw in politicians' deliberate attempts to manipulate our biases. Our blind spots get weaponized into a kind of autoimmune disease of the mind—the very mechanisms that evolved to make us wary of threats are commandeered to turn us into threatening (and ultimately self-threatening) mobs. "A reliable way to make people believe in falsehoods is frequent repetition, because familiarity is not easily distinguished from truth. Authoritarian institutions and marketers have always known this fact."

Kahneman's groundbreaking book was published in 2011. You'd think we'd have used his findings to be smarter consumers of news, to see more clearly, but social media companies—some of whose CEOs Kahneman led an exclusive master class for in 2007—and political leaders, many of whom had already intuited our cognitive blind spots, have been more successful at using the findings to manipulate us. In 2012 Facebook experimented with emotional priming on seven hundred thousand users; in the 2016 presidential election Russian bots flooded social media specifically to manipulate the biases Kahneman described. It's not a fair fight. As I discovered in the Vermont woods, the climates that surround us—whether acres of trees and snow or screens blizzarding with information—are more effective at changing our cognitive orientation than we are at changing it ourselves; memory, attention, and the way we make decisions depend more on the kinds of information we're surrounded by than on our own character.

In her 1967 essay "Truth and Politics," Hannah Arendt wrote:

> The result of a consistent and total substitution of lies for factual truth is not that the lies will now be accepted as truth, and the truth be defamed as lies, but that the sense by which we take our bearings in the real world—and the category of truth vs. falsehood is among the mental means to this end—is being destroyed.

That compass of truth vs. falsehood, that "sense by which we take our bearings in the real world," gets weaker online as we race through our feeds, not reading and thinking critically, and it stays weak as we cross back to our lives beyond the screen: in our conversations or lack of conversations, in snap judgements based on a hat or bumper sticker or t-shirt as we box each other into categories of us and them.

This is how the dominant "regime of the image," as Clark calls it, can aid and abet the coming of real-life regimes.

WHAT YOU CAN DO

When I told my friend Helena over the phone about my conversation with Steve, the A.R. friend, she said I was caught in the eighteenth century. Helena's a philosophy professor and one of the smartest people I know.

"Not even the twentieth?"

"What I mean is you're caught in an eighteenth-century question. We're all caught in it now. We've got two eighteenth-century camps built into our thinking. The first camp, which is pulling ahead, is the Enlightenment camp."

I pictured archery with a row of kneeling boys shooting at a second row of boys with apples on their heads. Then I remembered that story was William Tell and not Isaac Newton. "Enlightenment camp: Reason conquers all?"

"Sort of. I'll send you something. Go to your computer." What she sent was a quote from the contemporary philosopher Jeremy Waldron:

> The Enlightenment was characterized by a burgeoning confidence in the human ability to make sense of the world, to grasp its regularities and fundamental principles, to predict its future, and to manipulate its powers for the benefit of mankind. After millennia of ignorance, terror, and superstition, cowering before forces it could neither understand nor

> control, mankind faced the prospect of being able at last to build a *human* world, a world in which it might feel safely and securely at home.

The idea, she explained to me, was that knowledge and reason, properly marshalled, would lead to a transparent natural and social world, a more "human" one. As a corrective to the Middle Ages and its cults, we could now assemble facts, and reason our way through them, and figure out how to live the best possible lives. The undercurrent of this new faith, she pointed out, was that we could control the world, master it.

"Then what happened?"

"The second camp started."

My little archers perked up again. "Which was?"

"As a corrective to the Enlightenment corrective, the Romantics came along with skepticism about the desirability or even the possibility of such mastery and control. So we get Kant and Rousseau and eventually your friend Keats and his 'truth of the imagination' and his reverence for wonder and the unknowable. This is the second camp we've got built into our thinking, but it has a hard time winning debates with the first."

"I know the feeling."

"You're a Romantic," she said.

After we hung up, I went for a walk. Clearly Helena was right: it's impossible to debate against Reason without admitting that Reason has been useful to your argument. Keats wouldn't stand much of a chance on the TED stage. But the problem wasn't with Reason, which, frankly, I'd always liked, as long it knew its place and didn't try to dominate the proceedings; the problem was this cross-breeding of Enlightenment values with capitalism. The biggest collectors and analyzers of facts were no longer scientists; they were businesses like Google and Facebook (not to mention companies like Cambridge Analytica). And they weren't harvesting our data to help us figure out how to lead the best possible lives but to help their companies figure out how to make the greatest possible profit. That was the tangle. Facebook's abilities to gather facts would astound any Enlightenment philosopher, but the way Facebook analyzes and uses those facts leads to targeted advertising and newsfeeds that draw billions of people largely further away from facts, further away from reason, and, at least when it comes to politics, back towards the cultish ignorance the Enlightenment thinkers sought to end.

When I got back home, I called Helena.

"Of course," she said. "Facts are being harnessed for money—not knowledge or wisdom. Tech companies co-opted the Enlightenment camp for their own ends. That contradiction only complicates the debate between the camps. We need more clarity about facts when it comes to the news. And we need more reason when discussing those facts. But perhaps in other areas of our lives we also need more respect for the Romantic camp—for mystery, for instance, and for the possibility that some things should be left out of our individual or collective control."

"You're not just saying that?"

"I'm not just saying that."

We talked about how to get the two camps to coexist, how the two camps could support and enhance each other. Why couldn't facts and reason lead to more respect for mystery and humility? And why couldn't mystery and humility lead to the search for more facts and clearer reason?

ONE AFTERNOON AT THAT ARTISTS' residency in upstate New York, I was on the porch looking out at the lake when Oliver emerged from his studio to get his bathing suit off the railing. He looked enraptured, like he wanted to drink in the ferns and the lake in one great swallow of his eyes. "I've just been reading about earthworms!"

"Earthworms?"

"Do you know about earthworms?"

"Very little."

"Remarkable creatures! Did you know they have light perception?" His excitement cut the humidity in the air. The towels on the porch railing suddenly looked like banners of exploration.

"I didn't know that."

"They have no eyes, of course, but they do have photosensitive cells along their backs and sides. They know to avoid light, which signals to them that they're exposed to predators. Darwin's final book was on earthworms."

"That's what you were reading?"

He nodded like a proud child.

"What's it called?"

"*The Formation of Vegetable Mould Through the Action of Worms*!" he declared.

"Sounds like a Hollywood blockbuster."

Oliver said nothing. He was lost in reverie. The recently mown grass on the far side of the porch, the hemlocks by the lake, the hills going blue in the distance—everything seemed to have grown more wondrous to him. Perhaps he was imagining the millions of earthworms at work in the soil below us. Perhaps he was wondering at the unfathomably ubiquitous work of evolution, how it was in every glance on both sides of the eye. Or maybe he was just thinking about his beloved Darwin, a man who left the world a new understanding of the history of every species by paying close attention to a few species—and to his own observations—with reason and wonder.

IN THE FALL OF 2017, just two years after his TED talk about "experiencing the world better," iPod and iPhone designer Tony Fadell had this to say in a talk at the Design Museum in London:

> I wake up in cold sweats every so often thinking, What did we bring to the world? Did we really bring a nuclear bomb with information that can—like we see with fake news—blow

> up people's brains and reprogram them? Or did we bring light to people who never had information, who can now be empowered?

He didn't credit the 2016 presidential election or our growing national polarization for his change of heart. His explanation was more candid and personal. He had kids. He began to notice his children's eyes didn't want to leave their phones; they didn't like Daddy much, even though he was the god who'd created their phones, when he took the phones away. He said the phones weren't designed for what's best for families or the larger community—they were designed for self-absorption.

> A lot of the designers and coders who were in their 20s when we were creating these things didn't have kids. Now they have kids. And they see what's going on, and they say, 'Wait a second.' And they start to rethink their design decisions.

Which sounds promising: design flaws fixed through better design. The list of these digital regretters and reformers is now legion. They're like the robber barons who built great lodges in the Adirondacks to keep their families away from the throbbing metropolises they had grown rich creating. And now they're saying they want to build lodges for all of us. Credit to Tony Fadell for his candor, his conscience, and his hope.

But why would we look to Silicon Valley for wisdom?

Granted, in Plato's allegory of the cave, it's the captors who have to drag the prisoner out into the light. Behind the prisoners and unknown to them, the cave slopes upwards and a low walkway runs perpendicular to the entrance of the cave; unnamed powerful people walk along the walkway doing a kind of puppet show, the fire behind them casting shadows on the cave's innermost wall, which is all the chained men see. But when the captors finally release a man, the sight of the fire is too bright for him, he's dreadfully confused by the powerful people and the walkway, and he wants to go back into the heart of the cave. He has to be dragged outside, the daylight is blinding, and his eyes take a while to adjust: first he's able to see shadows on the ground (which must have been somewhat familiar to him, a kind of native language from the cave wall), then his eyes adjust enough for him to be able to see reflections of people in water, and then, eventually, he can see people themselves.

But Plato doesn't stop there. Just being outside is nowhere near sufficient for the man to see clearly.

The man begins to look at the night sky—he wants to orient himself, to get a sense of where he is. After more time has passed, he's able to glance at the sun, to feel oriented in the daytime too. In other words, seeing well takes more than being free of illusions, more than seeing beyond shadows. It takes seeing other people, and the heavens, and using all of that as a way of making sense of where you fit in.

Those large impulses towards orientation have nothing to do with anything the man's former captors can provide.

MISTY MORNING, long puddles holding a gray sky.

On the golf course, the fog is so thick I can't see the trees, the houses. About midway to the green, which isn't green at all, my neighbor John's house emerges. The usually red door the same black as the windows. A house just the shape of a house, a shelter with the low definition of a fairy-tale shelter for a family against the coming rain. The trees fill in as I walk closer, the lines of their branches, the door a smudge of orange, then warming into red. A few steps closer and the house blinks back into John's nice, ordinary house again, every line in place, a house where the mailman will soon deliver mail.

As I step back onto the street, it strikes me that my morning walks are like Post-it Notes to myself, the kind my mom put on my bathroom mirror when I was a kid. Unlike hers, they convey no information—no went to Star Market, no clean up your room. They just say: look around, go slowly, feel yourself a part of something bigger than yourself. Such a Post-it on my bathroom mirror would make me cringe. But to write and read it by walking just feels practical. Without the walk, I feel as though I've forgotten to do something important, something without which the rest of my day will go sideways. At bottom, maybe that's what prayer is. A kind of note to yourself, with your God looking on, written in your religion's handwriting, to remind you of whatever you think you need to be reminded of—to remind you how to live a better life. Maybe all the major religions knew that how you started your day would effectively be a prayer anyway, that whatever you do every morning, whatever frame you give to your day, effectively becomes what you worship. Even if you call it routine instead of ritual, there is a reminder to yourself there every morning, there is a frame that becomes an orientation, whether it's one you want to be living by or not.

Then I was at my front door.

JUST A FEW SUNDAYS AGO, on Harvard Avenue, I saw a teenager, maybe fifteen or sixteen, who had found her answer. She was with her parents. They were short and sturdy-looking, trudging along the sidewalk at a determined healthy pace, but the girl was jogging. At each side street she would turn off, vanish for a while, then return a few minutes later and jog in place alongside them, then come to another side street and jog off again. The girl's excursions gradually grew longer. Her parents didn't pay much attention to her, just seemed pleased at each check-in. I crossed to the parents' side of the street so I could see the girl's face. I wondered if she was annoyed at having to check in, if she felt like a dog testing its leash. But I couldn't have been more wrong. Each time she returned, her face was radiant. A full beautiful smile—and then she bounded off again.

Was it the pleasure of what she was discovering on her own down the side streets? Or the pleasure of returning and finding her parents just where she expected them? Or the pleasure of alternating between the two? I don't know. But she'd found her own balance between exploration and familiarity, her own rhythm between venturing off into the unknown and returning with what she'd found—a personal portal between frames which had become part of her own frame for experiencing the world.

She knew how to travel a border.

CHAPTER 4

May I Have Your Attention, Please

ENDANGERED TRAITS

Attention span (Attention from Latin *attendere*, *ad+tendere*, to stretch towards; and Span from Old English, *span*, to stretch)

Curiosity (from Latin *curiosus*, careful, inquisitive, from *cura*, cure)

Negative capability (Negative from Latin *negare*, to deny; and Capability from Latin *capere*, to take)

BACKGROUND

There was thin smoke, an acrid smell. Large shards and fluids of various kinds made a strange geography between our cars. She could walk. I could walk. That was all I noticed at first. This was Allston, the usually harmless student ghetto near Boston University. Three-story apartment houses, bicycles on front porches, the late morning tousled and quiet, sleeping off another Saturday night. I'd driven this shortcut hundreds of times, never seen a car come from where she came. Then I noticed the parked cars: on both sides of the street she'd shot out from, they faced the opposite direction.

"There was no stop sign," the young woman said. She pointed to the corner. "See, there was no stop sign. Google told me to come this way. It's not my fault. There was no stop sign."

My voice felt far away, as though it had kept traveling without me. I tried to explain that she'd come down a one-way street the wrong way, that stop signs aren't posted facing away from the direction cars are supposed to travel.

"What's wrong? You seem stressed, sweetie. I work for a lawyer, I know all about these things. Insurance will cover it. Don't be stressed." She reached out and touched my shoulder, which actually helped, though hearing her call me sweetie didn't.

The firemen came, we answered questions. The policemen came, we answered questions. She doubled down: "I'm going to sue Google. Why isn't there a stop sign?" The firemen swept the detritus, sprinkled down the fluids. The whole time, she didn't let go of her phone. That's where the story was. The one-way street, our mangled cars, her stiff neck, which she absently rubbed—they weren't as real as what was on her screen.

"EVERYONE," William James wrote in 1890, "knows what attention is. It is the taking possession by the mind, in clear and vivid form, of one out of what seem several simultaneously possible objects or trains of thought." The taxonomy of attention James came up with in 1890 is roughly what neuroscientists and neuropsychologists still use today. Attention requires selection—focusing on this rather than on that—and the focus can either be on something external (on an object, like a one-way-street sign or a smartphone, perceived through the senses) or on something internal (on a train of thought, like, did my boyfriend just text me?). Today's researchers designate these two branches of attention *external* and *internal*; James called them *sensorial* and *intellectual*. The other key distinction isn't between targets of attention but between the catalysts for it. Either your attention is voluntary, goal-directed, e.g., you want to check Google maps or for your boyfriend's text, so you look at your phone; or it's passive, stimulus-driven, e.g., a car darts into your peripheral vision, so your eyes dart up at the road. James called the first *voluntary* and the second *passive*; today's researchers call the first *endogenous* and the second *exogenous.*

According to this taxonomy—sticking with James, whose terms are more intuitive—our accident report is pretty straightforward. The young woman who T-boned me was likely caught in a rapid interplay of attention, glancing down at her phone (*voluntary sensorial*), absorbing a text or directions (*passive sensorial*), and perhaps texting back (*voluntary intellectual*), so she had no attentional capacity left over to look for a street sign (*voluntary sensorial*), or to notice all the cars parked in the opposite direction (*passive sensorial*), or to see the intersection with the main street ahead (*any kind of attention, please!*). So by the time my blue Honda flashed into her vision, it was too late. In taxonomical terms, we were screwed.

A few days later Andrew said, "It's like one of your articles crashed into you," by which he meant both *weird shitty luck* and *you should write about it.* "You know, don't text and drive. Don't look at your GPS and drive. Pay attention to the road." He was right, of course, but stiffness had rooted through my neck into my shoulder; a yellow and purple bruise had mushroomed along my thigh. I wasn't really interested in statistics on distraction (about one-quarter of car accidents in the US are tied to texting; the average time it takes to read a text is five seconds, which means if you're driving 55 mph, you drive the length of a football field effectively blind). And even if the woman's GPS had acted as accomplice, those stories were already so common they read as farce. *A Japanese tourist trying to reach an island off the coast of Australia drives straight into the Pacific. A driver in Belgium trying to find a city two hours away finds herself, two days later, in Croatia.* What concerned me, what really disturbed me, was how strange my conversation with the woman had been. She'd looked at me, she'd even reached out and touched me, but it didn't feel like she was there on the street with me. She couldn't take in her surroundings. Maybe she was just in shock, but I don't think so. She was pissed about the ticket the cop gave her. She wasn't pleased about the state of her car. But even when she wasn't looking at her phone, her attention still seemed geared for her phone, unable to go slowly enough to register the basic facts of what had happened.

That shift between the world on her screen and the world around her was what worried me—the 2,617 taps or swipes the average American makes per day. What did all those daily recalibrations between screen and physical world do to your attention? And what happened when you didn't adjust—when you were still using screen attention for the real world? And what exactly was screen attention—how did where you put your attention influence the kind of attention you paid? Was there a kind of attention exchange rate, so that using your phone changed your attention when you were off your phone? Was that why people kept talking about diminishing attention spans, was that part of why phones were so addictive?

Really, what worried me were the daily invisible attention collisions that happened without the headlines: the islands we weren't reaching in each other and ourselves, the whole countries we sometimes weren't noticing, the invisible crashes we only felt indirectly, after thousands of crashes had piled up.

This is what really frightened me. The kind of attention you rely on, and the kind of attention you let atrophy, ultimately becomes a decision about what you value as worth perceiving. "Each of us," James writes, "literally *chooses*, by his ways of attending to things, what sort of a universe he shall appear to himself to inhabit."

WHY IT MATTERS

My little black rental smelled like most rentals—half new car, half dusty motel room—and driving it made me feel as though I'd been forced to vacation in my own neighborhood. Three days after the accident, at the intersection at the end of my street,

I realized I was becoming a bad driver. A hypervigilant crash fantasist: the Coca-Cola truck that would run the red light; the cab that would swerve into the cyclist in the bike lane; the station wagon that would ram me. It was still hard to turn my head. I didn't trust other drivers to do what was probable, which meant I couldn't trust my own instincts, my sense of timing. In the growing dusk, I was driving in an anxious subjunctive: he might, she might, they might. I didn't know, waiting at the intersection, when to hit the gas.

The feeling reminded me of how, for a brief span when I was twelve, I resolved to be *always ready* to take the basketball court. I imagined that NBA players always had their bodies primed, their game faces on, so I thought this would be a good thing to practice. Walking the crowded corridors at school, going up the carpeted stairs at home, I kept my legs ready to cut for a pass that might suddenly appear, to leap for a rebound that might suddenly fall. It was probably a way of not having to pay attention to much else beyond my daydreams, or of paying attention only in the context of a daydream, an ongoing basketball game in which I was the star. It was exhausting, of course, and when I finally heard Larry Bird mentioning in an interview how he'd relaxed the night before a playoff game, it came as a relief. To play with Bird's otherworldly attention, his way of sensing passing lanes through a high-speed muddle of legs and arms, meant saving that kind of attention for the game. It meant he didn't walk through a hotel lobby imagining he was cutting through a defense. He just walked through the hotel lobby.

Near the market the traffic thinned out. Usually I enjoyed the drive, watching the taillights glowing red, electric sky fading to dusk. Usually I drove with a kind of half-attention suitable for the trees going to silhouette, for things half-revealed, for loose contemplation. But now I was like my kid's version of Larry Bird, overly ready, unable to save the right kind of attention for the right moment. I was hunting out threats at every cross street, giving the little fevered guards in my brain no chance at a cigarette break, and so I didn't see a thing.

In the market parking lot, accompanied by the cool breeze and a moody sky, I had a strange thought: when the young woman's car hit mine, maybe I'd absorbed her kind of attention—the anxiety of checking her screen, the need to be always in control—and now, when I drove, that kind of attention returned.

But maybe the young woman felt the same way about all the texts and updates crashing into her phone. On the morning of our accident, maybe she'd felt trapped in a kind of thermodynamics of attention. Maybe a few weeks earlier she'd been slow to respond to her boyfriend's text. Maybe the textbook progression had ensued: a second text asking for an immediate reply, anxious assumptions about the meaning of the lack of that reply, escalation of demand and urgency, culminating in unfounded and hurtful accusations. What was she to do? Her social responsibility now seemed to be not being aware of where she was but signaling where she was; it wasn't seeing what was around her but making herself visible to those who weren't around her. So, trying to be a decent person, she kept her texting and social media windows open more and more, and the attention climate outside, with its demanding expectations of speed of reply, kept blowing into her car. If she tried to keep her windows closed, her attention climate-controlled to her personal taste, she'd court accusations of being socially irresponsible and likely worse.

One unfortunate side effect of all this, of course, is that sometimes the old social responsibility crashes into the new.

BUT THE INVISIBLE collisions.

A couple of weeks after the accident, a student in my class delivered a Shakespeare sonnet as his poetry recitation. This was in Comp 101, a mandatory freshman English course at a college in Boston geared towards "makers"—by which the college website meant contractors, computer engineers, and civil engineers, not poets. Convincing the class that poetry had "applications" in their lives had been hard enough. But thanks to Robert Hayden and Seamus Heaney and Adrienne Rich, the students were now more or less game, or resigned, and, by their choosing, adventuring into the wilds of our basement classroom had come a raven, a noiseless patient spider, a man stopping by woods on a snowy evening, and a certain slant of light.

Kevin had signed up near the bottom of the list. He wore his ski hat as though snow might begin to fall from the paneled ceiling, kept one hand manfully wedged in his jeans, but his voice, when it emerged from behind a series of strange halting coughs, was as watchful as a deer stepping out from behind a tree:

That time of year thou mayst in me behold
When yellow leaves, or none, or few, do hang
Upon those boughs which shake against the cold,
Bare ruin'd choirs, where late the sweet birds sang.
In me thou see'st the twilight of such day
As after sunset fadeth in the west,
Which by and by black night doth take away,
Death's second self, that seals up all in rest.
In me thou see'st the glowing of such fire
That on the ashes of his youth doth lie,
As the death-bed whereon it must expire,
Consum'd with that which it was nourish'd by.
This thou perceiv'st, which makes thy love more strong,
To love that well which thou must leave ere long.

After finishing, Kevin gave a quick, effective synopsis of the poem, analyzed how it made use of the conventions of the Elizabethan sonnet, then began to address the question of why he'd chosen the poem, why it was meaningful to him. The symbol of the tree, the symbol of the birds, he read from his index cards, they are symbols of mortality. I hadn't once used the word symbol in class, and the clarity in his voice that had been so moving during the recitation had retreated behind what sounded like Cliffs Notes. I asked if he would keep going, just not using the word symbol or symbolize. His fist went to his mouth, he did his strange halting cough again, said he really hadn't slept that well last night. But then he said, "I'm from Maine, a small town outside Bangor. And we got trees all around, and I know how that looks, a tree in November or December, and how the branches look when the birds have gone, you know? Is this what you want?" I told him he was doing fine. "So, you know, I know how that looks, and how it feels when you're out with your dog or out there hunting, especially late in the day, like he says, when the sun's going down." He wiped his hand over his forehead, surprised, it seemed, by the hat coming off in his hand. "So we read that poem about the guy's father whose hands are cracked from all the manual labor he does, you know, and how he never thanked him." His voice was starting to go funny, and he was looking at me now like a deer who was trapped, who didn't want to go any further.

I told him I understood, told him he could stop if he wanted to, he'd done an excellent job. He swiped at his nose, picked up his index cards. But then he said, "How does he do that?"

"Shakespeare?"

"I mean what kind of guy is walking around paying attention to that stuff, like yellow leaves or none or few, the way the different trees are like that. And, you know, how different people are like that? How old men are like that?"

I thought of my father, the way a remaining strand of hair would sometimes lift ghostly in the wind. The classroom tilted in the silence.

I said to Kevin it seemed he'd been paying attention to all those things too—the trees where he lived, the way the light looked late in the day.

A couple of students laughed.

"So?" he said.

Kevin's voice was challenging now, unsure if this was a good or bad thing, if perhaps it was a kind of weakness. He'd mentioned in an essay how his older brother sometimes ragged on him for being daydreamy, for being a bad worker, and how ticked off it made him. So reaching for a frame of reference that wasn't personal, I told them about John Keats, how when he was about their age and thinking about Shakespeare, he'd written a letter to his brother saying that this quality of attention, of being open and observant, and of being comfortable with what you didn't or couldn't know, was exactly what was necessary for making "a man of achievement." Keats called it "negative capability," and the key was being "capable of being in uncertainties, Mysteries, doubts, without any irritable reaching after fact & reason."

"Just going out hunting and seeing what you see," Kevin said. "And not always having to have the answer for everything."

With the last part, I wondered if he was talking about his dad, or whoever it was that was sick. "Right," I said.

He looked down at his index cards. "And just spending time with people that way."

"Yes," I said, though it was something I hadn't really thought about with my own dad—how to be capable of being in uncertainties, how to see the few leaves hanging on the tree and not want to chase after the fallen ones; not, as he's catching his breath to finish a sentence, run through a kind of Google in my mind for why his cough has returned. How to enjoy time with him, comfortable with more and more unknowns.

Kevin put his hat back on and returned to his seat.

MAYBE THE VARIETIES of attention need better PR. In articles, the word "attention" is generally followed by the word "span" and dismaying information about how the reporter no longer has one and/or studies that prove that goldfish are racing—can goldfish race?—ahead of us in this capacity. "Attention span" seems the right place to start, but the articles rarely explain why. Occasionally, there's a reference to efficiency, to how long reading an article or report used to take the reporter versus how long it takes him now. That may be the most straightforward argument, but it's hardly the strongest or most important. What is an attention span made of? And why, apart from being able to read an article in one sitting rather than three, is it a valuable part of who we are?

William James admits that sustained attention is nearly impossible in and of itself. "No one can possibly attend continuously to an object that does not change." This is true of sensorial attention and "how much more true," James writes, of intellectual attention. But that change doesn't need to come from the object of attention itself. He quotes the German scientist and philosopher Herman Von Helmholtz on sensorial attention:

> If we wish to keep it upon one and the same object, we must seek constantly to find out something new about the latter, especially if other powerful impressions are attracting us away. . . . But we can set ourselves new questions about the object, so that a new interest in it arises, and then the attention will remain riveted.

If you're looking at an oak or a mulberry in Stratford-upon-Avon, for instance, the more you ask yourself about the pattern of the ragged remaining leaves, about the boughs shaking against the cold, about where the birds and their songs have gone, the longer you'll be able to pay attention to it.

Likewise, James writes of intellectual attention, "The . . . *sine qua non* of sustained attention to a given topic of thought is that we should roll it over and over incessantly and consider different aspects and relations of it in turn." The more you wonder about the ravages of old age in relation to the boughs of the tree, in relation to the vanished birds, the longer you'll be able to consider the ramifications—quite literally, the branchings—of old age. Setting yourself new questions, rolling over a topic of thought, considering different aspects of it allows flashes of insight, notifications from some terra incognita of your mind that you can't quite contact but that wants to contact you, to keep arriving. Attention span isn't a matter of willpower, or even necessarily of intellect. Attention span comes from curiosity! You focus your senses or your mind or both because there's something you're wondering at, or questioning, or ruminating on.

JAMES NOTES IT'S EASIER to define what strengthens attention than "to give practical directions for bringing it about," but he does suggest that teachers should "if possible awaken curiosity, so that the new thing shall seem to come as an answer, or part of an answer, to a question pre-existing in [the] mind." Writers have known this for a long time—a cliff-hanger is nothing if not a question: Will Juliet wake up before Romeo kills himself? Will Othello learn the truth about Desdemona before it's too late? Will Macbeth ever find strong-enough hand soap? But those questions, if the writing is good, depend on you caring about the characters. Clickbait headlines skip the caring and go straight to questions, or top-ten lists, so you can't help but feel the responses are to questions you've been considering yourself: just what *were* the top ten shower/horror/sex/romantic scenes of all time? Suddenly you're saddled with a prefab curiosity about a question you never really cared about.

Perhaps the most insidious version of this prefab curiosity is the YouTube autoplay algorithm I mentioned earlier. "What keeps people glued to YouTube?" Zeynep Tufekci wrote in a *New York Times* opinion piece. "Its algorithm seems to have concluded that people are drawn to content that is more extreme than what they started with—or to incendiary content in general." So as the next more adrenaline-inducing video loads, the autoplay circle fills as though guiding you one level lower into some great mystery—down into the frightening truth of hoaxes and conspiracy theories. Even just watching ski videos, which I do more than I should, leads me more rapidly than my ski imagination is ready for to an extreme descent through the most dangerous couloirs in the world. I don't start with a hankering for death-defying stunts, just as most people don't start with a hankering for political extremes. But YouTube plays on our natural curiosity—to uncover more and more hidden answers, more and more dangerous descents—and dulls our curiosity itself, constantly asking the next question for us, and usually fairly dumb thrill-seeking questions at that.

While not trending towards extremes, and so far less pernicious, every major media outlet works a similar curiosity angle. The "recommended for you" articles at the bottom of each article, if you're using the *New York Times* app, might be answers to excellent questions, might be considering important angles and implications, might even ask a smarter question than the one you would have asked yourself. The same goes for Netflix and other entertainment apps when recommending your next movie; the same for commercial apps when recommending your next purchase. But in every case your curiosity guide, whether an algorithm or a digital editor, sweeps in like a fairy godmother, and sooner or later becomes your curiosity surrogate. There's no need to formulate the next question, angle, or association for yourself. You just have to click yes, that's what I wanted to ask, yes, that question, without realizing that while your attention is being engaged, the engine of your attention, which is your own questions, is not. There's no need to move from consideration of an article or movie to a kind of investigative logic, to a next step in the progression of what you want to learn or see.

Curiosity as an approach to the world, as a means of orientation, is becoming obsolete. I don't just mean Google curiosity—with questions that can be instantly searched, instantly answered—or online news curiosity—with questions that get asked for you—but the kind of curiosity that originates with negative capability, with following your deepest affinities, the kind of curiosity that carries you branch to branch. Perhaps that's why two-thirds of men and one-quarter of women would rather receive a shock of *electric current*—a 2014 study actually found this—than sit alone with their thoughts. Being bereft of questions leads to a maze of boredom, with no curiosity to lead you through. What are the top-ten most popular thoughts to have next? Why do intriguing follow-up links not appear?

William James writes of geniuses, "Every subject branches infinitely before their fertile minds, and so for hours they may be rapt." Digital life can mimic this feeling of being a genius—with link after link, question after question, a subject branching infinitely—but once off-line we lose our curiosity superpowers, and often feel a dazed emptiness of return: we're alone in our rooms, we haven't been understanding anything of significance, we're not geniuses at all.

For what, asks the goldfish watching us from the bowl, have we traded our attention spans?

THREATS

The pasture where a teenage boy tends his master's sheep is far enough from the village that no one comes to visit, yet close enough that he can feel village life humming along without him. His only amusements are talking to his faithful mutt and playing his pipe. Bored, agitated, lonely, he daydreams about the dark forest nearby. One afternoon the daydreaming gives him an idea. He stands from the pasture and runs down the hillside shouting, "Wolf! Wolf!" The villagers drop their work, rush up the hill, only to find the boy's sheep quietly browsing the dry grass, and the boy laughing. A few days later, the same running and shouting boy, the same helpful villagers, the same browsing sheep and laughing boy—and the same pissed-off villagers, though more red-faced. A few evenings later a wolf creeps out from the forest shadows and falls upon the sheep. Same running and shouting boy, no helpful villagers. They know better.

But imagine if the villagers grew disappointed with the boy who cried wolf for not crying wolf enough—for not giving them the chance to come racing to see if the sheep were safe, for not giving them that charge of excitement. Our forty-thousand-year-old brains have basically turned into those villagers. Having heard our screens cry wolf so often, our brains don't learn not to heed the call; they learn to love it. A wolf could be there! Or, better yet, someone reaching out for us could be there—a joke, a flirtation, a kind word. To our brains, the potential for those things is almost as good as the thing itself, because it has a similar effect: something to get excited about! Dopamine gets released with each notification, and our brains have no satiety system for dopamine, nothing that says we've had enough. We evolved to be highly receptive to sensory stimuli—a snapping branch, a flash through the trees, a sudden change in the shadows—which means we evolved to be highly distractible, because our survival depended on it. But now flashing notifications trip our primal alarm systems, alert us both to the possibility of danger and of potential reward, and then pull us into a feedback loop of wanting more alarms, of going *searching* for them, to release more dopamine.

Part of why the digital boy who cried wolf is so successful is because of what psychologists call *variable reward*. Since the messages and likes don't always come, you're driven to click and tap to make them come. Nir Eyal, a graduate of the Orwellian-sounding Persuasive Technology Lab at Stanford, and the author of *Hooked: How to Build Habit-Forming Products*, explains in his blog post "Want to Hook Your Users? Drive Them Crazy":

> At the heart of the Hook Model is a powerful cognitive quirk described by B. F. Skinner in the 1950s, called a variable schedule of rewards. Skinner observed that lab mice responded most voraciously to random rewards. The mice would press a lever and sometimes they'd get a small treat, other times a large treat, and other times nothing at all. Unlike the mice that received the same treat every time, the mice that received variable rewards seemed to press the lever compulsively.

This is why it's hard to resist a slot machine. Every time you pull the lever or push the button, the cherries might line up, or they might not. On an evolutionary level, it's like being able to control when a wolf, or the chance of a wolf, will bound out of the underbrush. Or, better yet, when a deer will bound out of the underbrush. You can control when a deer will bound out of the underbrush! You want surprise, you want reward, but you don't want the waiting. Deer-on-demand! Drive-thru hunting! You might not get the biggest buck, but something's bound to be flushed! And if a wolf's out there, you'll know that too! Or you can always try again, again, and again. You can control when you need to pay attention. But the irony is that you start to lose control over that control: the trying becomes compulsive, and now you're paying attention all the time.

Now turn those wolves/deer/cherries into social approval, add the ability to grant other people wolves/deer/cherries, add the ability to compare your quarry of wolves/deer/cherries to everybody else's, and 2,617 touches a day doesn't sound so far-fetched. As Tristan Harris, a Persuasive Technology Lab dropout and a former engineer at Google, wrote on his digital reform blog: "Just like the food industry manipulates our innate biases for salt, sugar and fat with perfectly engineered combinations, the tech industry bulldozes our innate biases for Social Reciprocity (we're built to get back to others), Social Approval (we're built to care what others think of us), Social Comparison (we're built to care how we're doing with respect to our peers) and Novelty-seeking (we're built to seek surprises over the predictable)." The last part is the real kicker. Silicon Valley's young engineers are designing the perfect tools to manipulate our attentional instincts in order to manipulate our social instincts.

Casinos make 80 percent of their revenue from slot machines. They're called one-arm bandits because the original machines, designed with a single lever on the side, often bankrupted people.

ONE OF THE GREATEST GAMBLES when you're young is leaving town. Packing up the car, checking the map, feeling the land open to you, feeling the possibility that you'll find more of yourself out there on the horizon, with a new sky and new opportunities, with a different scent in the air and a different light. Before moving to the woods I spent two years living for short stints in New Mexico, Idaho, and Montana, and in communities along the way, I found a smaller kind of gamble that was part of that larger one, communities where you could barter. This happened with artists and farmers mostly in leftover hippie enclaves: a poem for a charcoal sketch; a few hours mucking out a barn for a week of fresh tomatoes and basil and garlic. You knew what you were giving and what you were getting—the only gamble was what the work might reveal to you about yourself. After the big trade of leaving Boston for parts unknown, the small trades were a satisfying comfort.

But now when I barter my attention online, give it for twenty minutes to Facebook, calculating the worth of the trade is impossible. Even if I can estimate the short-term benefit of spending that time with my newsfeed, I don't know how my likes and dislikes will return to me through the stories and behavioral advertising I get fed, and I don't know how my attention itself will have changed when I look up from the screen.

Thanks to the work of Cambridge Analytica during the 2016 presidential election, the data part of the trade is getting media attention, and grudging contrition from Facebook, and may even get some legislation. But that doesn't help with my attention exchange rate. The more attention I give my newsfeed, the more rapid I expect the interplay between *voluntary* and *passive attention* to be. And of course this doesn't just happen on Facebook. Nearly every app caters to the urge for control (*voluntary attention*)—click or tap wherever you want—while also catering to the instinct for change (*passive attention*)—who knows what might pop up? The interplay between the two occurs at a highly stimulating rate, unparalleled in life off-screen, except perhaps in moments of anxiety. (Can you imagine your eyes darting around the room—look/absorb, look/absorb, look/absorb—as rapidly as you click or tap?) Life's in-between moments, like waiting in line, or driving to the market, or going for a walk, depend on a relatively slow interplay: your *passive attention* waits to see, or hear, or think something out of the ordinary, you engage with it, your passive attention waits anew. Online, that kind of waiting isn't necessary.

Meanwhile in just twenty minutes, the part of my attention that depends on patience—the patience to formulate a question, the patience to wait for an answer—has gone slack. I've lost more than my time and data. I've lost the sharpness of my curiosity, and also a less conspicuous kind of attention, a restful kind I'd come to love on the road, which is basically just one quiet question after another asked effortlessly by the senses and answered in a soothing stream. In New Mexico it was the rhythm of the yellow lines slipping past, the wide open spaces and big sky unfurling, no flashes other than the heat lightning, the land opening through mesas and arroyos and bluffs in slowly shifting geometries, the blue clouds inviting and heavy, the stopping at a roadside store just a continuation of the feeling. That was a trade on my attention, when I pulled into Chimayo or Arroyo Seco or Ojo Caliente, that had always improved my attention and made it easier to wait on the wild and quiet in the people I met.

My hunger for something I couldn't name transformed into a kind of patience on the road, a kind of attention forced on me by knowing so little and needing so much. What worries me now is how similar needs, in each of us but especially in young people, are being fed largely by the sights and sounds on screens, which ultimately weaken the kind of attention we need to absorb whatever it is we really need. The trade-offs between camouflage and display underlie the appearance of nearly everything in the natural world, including people (I'll always remember getting dressed for high school dances, the uneasiness of trying to look cool without being a target for insult). But apps have no need for camouflage—all their energies go towards display. The evolutionary advantage goes to the apps. How to pay attention to magpies and western bluebirds instead of your phone, or even to the person riding shotgun beside you, if they're not so perfectly likable, so constantly attracting your attention, so ready to do your bidding?

What worries me is not the computers-go-bad/conspire/take-over-the-world scenario; it's not even the people-designing-them-go-bad/conspire/take-over-the-world scenario. It's simply the we-no-longer-have-the-patience-to-pay-attention-to-the-world scenario, which includes, as dystopian sequels, the we-no-longer-pay-attention-to-each-other scenario and the we-no-longer-share-our-vulnerabilities-and-never-want-to-deal-with-anyone-else's-vulnerabilities scenario. What worries me, ultimately, is that we won't love. Not the world around us, not each other. We won't need to. Why wait for a deer, of any kind, to come bounding out of the trees?

MAYBE TEN DAYS AFTER THE CAR ACCIDENT, my brother and sister-in-law were in town and I was in the backseat, riding with my niece Emily, Sophie's older sister. The word precocious doesn't do her justice. In third grade for her class's Hero Day, long before the RBG t-shirts and movies and mugs, she dressed up as the Supreme Court justice, her black robes all the more heroic beside the miniature Iron Men. Now, at fifteen, she was smart, authoritative, and concerned her uncle was out of touch. She seemed to be having a conversation in her head as my brother and sister-in-law talked in the front seat. Then she turned to me. "Technology is progress, you know."

"Always?"

"Always."

"What if something already works well?"

"There's no invention that can't be improved."

I looked out the window for help. "How about the traffic light?"

"Could be better."

"How?"

She let out a breath. "I don't know, I'm not a traffic light designer. Someone will think of it."

Later that night, I checked Wikipedia. Traffic lightbulbs have been made more energy-efficient; visors have made them easier to see; computers have made signals more responsive to traffic patterns. Each of these features, especially the algorithms, will continue to be improved. My niece, not surprisingly, was right.

But Wikipedia also told me that the basic design of the traffic light has hardly changed since the 1920s. Three colored circles in vertical formation. Each color and location a clear indication: go; slow down; stop. Sometimes a fourth light with a turn arrow. It's a design created nearly a century ago, by a police officer in Detroit named William Potts, and it has saved countless lives every day.

The next morning, on the corner by the traffic light two blocks from my apartment, I couldn't help marveling. Everything on the street had changed around it—the cars, the storefronts, the fashions people wore. A century of shifting tastes, shifting mores, and the traffic-light design, which didn't look particularly exciting or efficient suspended there over the intersection, was largely unchanged since Woodrow Wilson was in office. But crossing the street, I realized the traffic-light design hasn't changed not just because it works but because drivers can't buy new traffic lights. If they could, then traffic-light companies would have to innovate to compete, not just for government contracts but for our hard-earned money, which would mean appealing to our sense of identity, our sense of style. Sleek, fancy traffic lights would pop up in wealthy neighborhoods, hip traffic lights in Williamsburg. Some nondescript Midwestern city would emblazon its quirky design on t-shirts and mugs, its new claim to fame. Each year's model would be touted as new and improved, capable of inspiring envy in other neighborhoods. The first priority for traffic-light companies would no longer be safety; it would be sales, which would mean attracting enough attention to sell. So, design flourishes, cross-pollination with fashion, traffic-light magazines for the aficionado. (*Signal Weekly* would be the joke of the industry.) The big time would follow: car crashes, class-action suits, quips and outrage on cable news, cartoons in the *New Yorker*.

For a few days, traffic lights became my city redwoods—it comforted me to think they had witnessed history streaming by, the days turning into decades. But what's most impressive and distinctive about Potts's traffic-light design isn't its longevity or even its safety transparency—how quickly we register its effect on our lives—but how perfectly it calls on our attention. It asks for only as much attention as it requires. It never cries wolf. If a traffic light blinks or changes, you should pay attention. In the attention economy, how many blinking lights can we say that about? App designers, not to mention cable news and movie directors, would likely design better products, or, equally important, might not be forced to design worse ones, if they didn't face such overwhelming pressure to win the attention war.

Thank God a traffic light, which needs to get our attention to keep us safe, doesn't need to compete for our attention in the marketplace. If it did, it likely wouldn't keep us safe at all.

WHAT YOU CAN DO

The guy from the body shop called to say my car was ready. He'd aired it out, he said, there was no more smell. When I opened the car door to a warm waft of lemony vomit, he said deadpan that after working so long with fumes, he no longer had a sense of smell.

For the next week, I tried to think of the car's fumes as a kind of personal warning of climate change, a dose proportional to the effect my car's emissions would eventually have on the environment. What if such warnings came standard? Power windows and a little toxicity to help you towards the greater good! It was easier to think this way when I wasn't actually in the car, of course, which had me driving with my nose and mouth buried in the collar of my jacket, like a child's version of a spy.

And what about an analogue for inner climate change, a personal warning that could come standard but never will? Maybe when you're tapping app to app, a doppelganger window in the top left corner of the screen could show you aspects of your life given the same level of attention you're giving the pages you're viewing—the details cursory, the larger context absent—and the top right corner of the screen could enact your attention-exchange rate, showing you the diminishing diversity of your attention: a highlight reel of the future moments you won't be attuned to, the intimate revelations and reflections you won't have the patience to wait for, the questions you won't have the curiosity to ask, all the unbidden glimpses of something larger than yourself you will not get.

FOMO, OR FEAR OF MISSING OUT, didn't start with a smartphone. It started with the human condition, and it wasn't only a social fear but also a practical and spiritual one. It likely arose from the fear of being left alone in the bush, and of not hearing or seeing a real threat lurking in the branches, and the broader, concomitant fear of not understanding how the world worked, its patterns and rules. We needed to pay attention to survive the world but also not to miss any clues to the mysteries behind the world: behind social relations, behind the thunder and the lightning, behind good fortune and misery.

There's a famous story of a man who went to seek advice from the fourteenth-century Zen master Ikkyu. When he asked for the highest wisdom, Ikkyu lowered his pen to parchment and inscribed the character "attention." The seeker, fatigued and somewhat insulted, said, "That's all?" The master picked up his pen again. "Attention. Attention," he wrote. "I really don't see much subtlety in that," the seeker said. "Attention. Attention. Attention," the master wrote.

Five centuries later, William James's son Willy would graduate from Harvard. Anxious about finding a job, paying taxes, knowing how to live and what to live by, he asked not his famous father but his famous uncle, the novelist Henry James, for advice. Henry answered, "Three things in human life are important. The first is to be kind. The second is to be kind. And the third is to be kind."

In some ways, it's the same answer. Kindness isn't just a habit of polite inquiry—it's an endless search, a constant Sisyphean effort towards a way of being with other people. Now apply that orientation not just to other people but to everything, everything, and you've got Ikkyu's answer. The implication of the repetition is that there are many kinds and many levels of attention, just as there many kinds and many levels of kindness, and that they're increasingly difficult to master. For Ikkyu, practicing attention with this in mind, and then practicing attention with nothing in mind, leads to the highest wisdom. But perhaps most telling is the seeker's response: his impatience with the first answer, which he takes as a personal insult; and his lack of curiosity with the second answer, saying he doesn't "see much subtlety" in Ikkyu's repetition, when curiosity about subtlety is exactly what Ikkyu's repetition demands (whereas the seeker, we infer, only uses repetition for emphasis). The very kinds of attention Ikkyu implies are the highest wisdom—patience, the ability to remain in mystery and not reach after reason, and curiosity—are the very kinds of attention you need to contemplate his answer, the very kinds of attention the seeker does not have.

My fear is that digital life, as guru, transforms us into this kind of seeker. Rapid interplay between voluntary and passive attention can be extremely useful, even necessary at times, but for that hybrid species of attention to displace so many others, for it to invade territories in which it's not only non-native but clearly hostile—driving, or talking to a friend, or daydreaming, or walking in the park—diminishes the rainforest diversity of our attentional lives, and so our range of experience. With 2,617 clicks a day, under the guise of experiencing more, we actually experience less. We roam the digital woods, chasing wolves and non-wolves, deer and non-deer, rarely getting to step outside the screen, to wait and watch as something we couldn't have imagined emerges, alert and wary and alive, from the trees.

I remember the first time I heard the story about young Willy James. It was at lunch with my thesis advisor, Bob Coles, a week before my college graduation. The restaurant had stained-glass windows, library-like dark wood paneling, and terrible food. I didn't want the meal to end. Coles was the wisest person I knew. The heart of his career had been traveling from New Orleans to Alaska talking to children from all walks of life about their moral and spiritual lives. As we waited for the bill, I took the chance to ask what I'd always been too shy to ask, which was if he had any advice for life after college.

Coles said he didn't. Maybe he sensed my disappointment, and my need, because after a moment he described young Willy James writing a letter to his uncle Henry with the same question, and then he gave Henry's answer.

"The first thing is to be kind."

Yes, I thought, hold doors, be polite, not a bad warm-up.

"The second thing is to be kind."

Yes, very important, but now for the punch line, the third man who walks into the bar, the real answer.

"The third thing is to be kind."

That's when my mind went quiet. The conversations in the booths behind us, the sounds of silverware on plates, everything slowed down. Each of the three answers started to arrive again, more slowly, as though from farther away, this time with enormous spaces in between them.

Like the starlight out my window, they're still arriving now.

CHAPTER 5

Identity Theft

ENDANGERED TRAITS

Spiritual self (Spiritual from Late Latin, *spiritualis*, from Latin, of breathing, of wind; and Self from Old English, *self*)

TWO YEARS AGO an English teacher at the very private, very all-boys high school I had attended emailed to ask if I would deliver a Hall. Hall is the morning convocation: the headmaster and speaker enter single file, the boys rise as one from rows of hard wooden seats; the headmaster and speaker mount the stage, and with a faint downward motion of the headmaster's hand, the boys sit. Having already delivered a series of Underground Man rants in my head against the school in my high school days, rants unworthy of Dostoevsky but not shy about leaning on quotes from him ("The whole meaning of human life can be summed up in the one statement that man only exists for the purpose of proving to himself every minute that he is a man and not an organ-stop!"), I felt honored and ashamed by the invitation, as though given a chance to revise a dream that had proven wildly myopic. The school had changed considerably in the twenty-five years since my graduation, keeping its strengths and doing away with most of its stuffy weaknesses, and I worried that arriving for a dream two decades late could only end badly.

But on the drive over, time had hardly passed—the same three-family triple-deckers, the same liquor store on the corner (advertising six-packs of Corona now instead of Bud), the same shamrock dotting the *i* of Corrib's Pub. The school had a new parking lot, a few new buildings, but inside yawned the same long corridors, hung the same austere portraits, and ricocheted the same Boston accents. There was even the same swarming pattern at the bell. My chemistry teacher approached, the same camera hung around his neck for his double role as yearbook advisor, hair grayed as though with baby powder for a school play about the school itself. After the reading, which was uneventful apart from my feeling that I was sitting out on those hard seats at the same time as I was standing up on stage, a friendly, bald man pumped my hand and said, "Remember me? It's your old friend Kent!" He had been a classmate; his son was now a student.

But then something strange happened, a hitch in the loop of time. For an interview for the school newspaper, *Tripod*, two boys escorted me to Room 12, the classroom near the parking lot where we used to wait humidly for carpool after basketball practice. They played good cop, bad cop. The short one, who had a wispy beard and a direct professional manner, asked questions about my years in solitude with subtexts of mental health and responsibility to my family; the lanky one nodded sympathetically, as though to say, sorry just doing our job. Late in the interview the short boy ventured something personal: "My parents give me a lot of grief about how much time I'm on my smartphone. What do you think?"

It was easy to imagine him as a successful journalist but just as easy to imagine him outside Room 12 by the water fountain, struggling to decode a group of teenage girls there for a dance, stroking his hope of a beard, having no idea what to say or who to be. His parents were probably worried less about their intelligent and vulnerable boy's attention span, or even how he'd come to know the world, than about how he would come to know himself.

I said that as a teenager I lobbied my parents for a TV in my room, and then for a phone, but got nowhere on both counts.

"Same fights, different details," the boy said.

"Kind of."

"So what'd you do at night to fall asleep?" the other boy said.

I said I'd lie in bed thinking about the moments in the day that had given me pause, moments that stayed with me probably because they contained an element of the mysteries I was trying to figure out.

"Mysteries?"

"My older brother, my girlfriend, why the basketball coach wouldn't give me more playing time, the green light in *The Great Gatsby*."

No trace of a smile—they really were little professionals. Hadn't they figured out that the green light was the symbol of all High School English Bullshit symbols? Fearing I'd lost them for good, I tried to explain how I'd reflect on moments and conversations from the day, and how there was something comforting about returning to that place in my mind night after night, about building up questions and the beginnings of answers, not because I usually got anywhere, but just because I was building up a sense of myself, a way of approaching questions that were important to me. And then I'd go to sleep.

The boys looked at me over their desks as though I'd told them a story about dragons and knights, some foggy and distant legend with no bearing on their daily lives except as hidden allegory. It chilled me to think that the place inside of me that seems the center of my identity, a place I didn't really know how to describe, might not correspond to any place inside of them. With all the changes the school had made, I never would have imagined the change that would strike me most would have nothing to do with the school.

The boy with the wispy beard motioned to the phone on the desk. "I do wish I could chuck it out the window sometimes. But it's a social necessity."

"That's just the way things are now," the other boy said. "Like it or not."

NO ONE ACTUALLY knows what the self is. There are just overlapping and conflicting theories: Sociological theories such as Erving Goffman's that the self is a performer, playing a series of roles to manage others' impressions, masks and costumes switched as the situation demands, with no authentic self waiting with flowers in the wings. Or Pierre Bourdieu's theory that the self is largely constituted by cultural and class distinctions of taste, so we like what we like not from any innate personal affinity but from an adherence to a kind of culturally preapproved menu, which we learn and internalize from an early age, liking country music or Shakespeare or kale smoothies as we're supposed to, thereby keeping different classes in their "proper" place. And there's the commonsense psychological theory, itself a kind of smoothie blending Freud and capitalism, that the self is basically a CEO overseeing all our operations, with lower-level insubordinate urges sometimes acting on their own, only to be reprimanded by the boss. Or evolutionary psychology theories that the self is just the interplay between competing impulses like attracting mates, fighting off rivals, securing social status, and caring for kin.

Any of these theories might have helped me to answer the boy's question, to talk about why maybe his parents were concerned about his screen time, "All the world's a stage"even more apt with social media, the performative self constantly performing and arranging performances, no rest walking home from school or lying in bed from the social pressure of being a teenager. Which leads to Bourdieu, the pressure on taste, trying to classify yourself as cool or uncool by what you classify as cool or uncool, the pressure no longer just on the band names inscribed on your notebook, or the primeval social media of your yearbook page, but compounding with every post, with every like or view or horrifying non-response, your social status vulnerable to public disaster at any moment. Which is where your forty-thousand-year-old brain comes in, your evolutionary impulse to secure and improve social status, so you don't get banished from your village, so you don't have to face the dark night alone in the bush.

But none of those versions of the self is what I wanted to explain to the boys. None of them touched on the elusive central part, the part I was just getting familiar with as a teenager at night in bed, the part I didn't really get to know until after my eye accident, retreating to it morning after morning looking out my dorm-room window, because the street and the sky looked different, and it was the only part of me that felt the same. It wasn't a particular thought or feeling but just the most simple point of contact between me and the world around me, a wordless, effortless receptivity to a passing pigeon or a passing thought, a kind of personal airport where perceptions or ideas came in for landing or they didn't, an airport whose runways would change sometimes depending on my interests or my mood or things I didn't understand, but whose land inside me was the same as it had been night after night in my bed as a teenager, a place with a distinctive atmosphere even when the weather changed, even when I drifted far away from myself, even when it was just a string of bleary runway lights through the fog.

A FEW MORNINGS AFTER talking with the boys, throwing out my lanyard from a literary conference, I noticed a poem towards the back of the conference's booklet, or maybe it was part of a poem; I couldn't tell.

> I glanced at her and took my glasses
> off—they were still singing . . . "I am your own
> way of looking at things," she said. "When
> you allow me to live with you, every
> glance at the world around you will be
> a sort of salvation."
>
> —"When I Met My Muse," by WILLIAM STAFFORD

Maybe this was what I should have said to the boys: How do find your own way of looking at things? How do you take off your socially preapproved glasses, so that in every glance at the world around you, you feel a part of the world around you, even saved by your sense of belonging in it—a belonging rooted not in your wearing the same glasses as everyone else, but precisely in having taken them off, in feeling your common solitariness with everyone?

Here was one room; there another.

An invitation to your own personal loneliness, further explained by a quote from Virginia Woolf? Teenage boys would love that—I was going from bad to worse. I needed help. So I checked to see if there was more to the poem, and found it on a site called Poemhunter.com. They had a video, I assumed of Stafford reading the poem, so I clicked. There was a still image of the poem's full text:

I glanced at her and took my glasses
off—they were still singing. They buzzed
like a locust on the coffee table and then
ceased. Her voice belled forth, and the
sunlight bent. I felt the ceiling arch, and
knew that nails up there took a new grip
on whatever they touched. "I am your own
way of looking at things," she said. "When
you allow me to live with you, every
glance at the world around you will be
a sort of salvation." And I took her hand.

Before I could finish reading, an automated woman's voice began. She was intoning the poem. She hadn't been updated in a long time, her voice stilted and tinny, like a voice run through a rusty strainer. The hairs on the back of my neck stood up when she said, "Iam your ownway ofseeingthings," and again when she said, "This willbe a sortof salvation."

MAYBE WILLIAM JAMES—who knew from personal experience, including the death of his two-year-old son Herman, how your sense of identity, however strong and stable it seemed, could give way under enough strain—could have been a help. In his *Principles of Psychology*, he divides the self into three parts. First, there's the material self: your body and its instincts, and, somewhat strangely, also your home and property. Then there's the social self, which is where he gets helpful. "A man's Social Self is the recognition which he gets from his mates," he begins, as though raising a glass at the bar, but within a few sentences he sounds as though he knows the despair that would drive a man to drink:

> No more fiendish punishment could be devised, were such a thing physically possible, than that one should be turned loose in society and remain absolutely unnoticed by all the members thereof. If no one turned round when we entered, answered when we spoke, or minded what we did, but if every person we met 'cut us dead,' and acted as if we were non-existing things, a kind of rage and impotent despair would ere long well up in us, from which the cruellest bodily tortures would be relief; for these would make us feel that, however bad might be our plight, we had not sunk to such a depth as to be unworthy of attention at all.

Such a "fiendish punishment" isn't physically possible, James implies, not if you're with other people in the flesh, or at least not to such an annihilating degree, but in an uncanny way, his words prefigure the threat of social media, the "fiendish punishment" of getting no views, the fear of being unworthy of attention cycling with brief fixes of attention that are never enough, not to mention the increased likelihood of being ignored, unworthy of attention, when you're off-line, surrounded by people who are on their phones.

This is where the third part of the self, the spiritual self, comes in to save the day, or should. James writes that it is the very center of identity, capable of anchoring us if our material and social selves are under siege, even worth sacrificing friends, reputation, property, and life for, if necessary. But he struggles to explain just what the spiritual self is—it's elusive to him too. Eventually, he calls the spiritual self a kind of gatekeeper:

> A spiritual something in him which seems to *go out* to meet these qualities [of his feelings] and contents [of his thoughts], whilst they seem to *come in* to be received by it. It is what welcomes or rejects. It presides over the perception of sensations, and by giving or withholding its assent it influences the movements they tend to arouse. It is the home of interest,—not the pleasant or the painful, not even pleasure or pain, as such, but that within us to which pleasure and pain, the pleasant and the painful, speak.

Gates and gatekeepers were far more common in James's time than airports, but basically it's the same idea: the spiritual self determines what we are receptive to. James doesn't say much, though, about where the spiritual self comes from, other than the suggestion that it is God-given, or whether it can be influenced by the social and material selves. Which is why I might have brought it up with the boys. What did they think? What determined what they were receptive to? Was any part of their spiritual selves not touched by their social selves?

Or had their inner gatekeepers lost their jobs to likes and views, to an internalized personal algorithm for what would make them more popular?

WHY IT MATTERS

My spiritual self didn't seem all that important to me in high school, more of a curiosity than anything else. True, without it,

I probably wouldn't have watched so much BET (my brother pointed out that it stood for Black Entertainment Television, and I was a white Jewish boy in the suburbs), or felt so moved by the smell of pinon smoke near the Grand Canyon and the spray of stars overhead, or felt such a closeted attraction to James Joyce's short stories, or listened to Bedtime Magic with David Allan Boucher (pronounced boo-*shay*), on Boston's easy listening station WMJX, long after I knew it was the furthest thing from cool.

Of course, Bourdieu fans and good old-fashioned cynics would argue that a spiritual self independent of a social self is largely a fantasy, that really I liked BET not because it was culturally approved but precisely because it wasn't, or that being moved by the smell of pinon smoke at the Grand Canyon was just another way of setting myself apart from my suburban family (even if only to myself), or that Joyce touched not a hidden love of language but a hidden self-regard, something inside me just waiting to be validated as different and special, and that my listening to easy listening once I knew it was uncool was just a way of trying to be cooler by saying I wasn't subject to the pressure to be cool.

Maybe. Maybe social pressures are atmospheric, and the center of the spiritual self, however deep inside, still breathes them in, whether you're aware of it or not.

And others would point out, rightly, that as a white boy I had a far easier time not seeing myself through other people's eyes, not growing up with that double-consciousness Du Bois wrote about in *The Souls of Black Folk*: "It is a peculiar sensation, this double-consciousness, this sense of always looking at one's self through the eyes of others, of measuring one's soul by the tape of a world that looks on in amused contempt and pity." So even if my affinities did come from some authentic spiritual self, that easy access to my spiritual self was itself a privilege of my social position as a white boy in the suburbs.

True.

And yet, everyone, whatever their background, has a material self, a social self, and a spiritual self, even if the pressures among those selves vary enormously depending on their background. And everyone, at some point, after some trauma or loss or dark night of the soul, will wake to an uncertain morning when not even their bedroom floor feels solid. At least, that's what happened to me. And though I leaned on my family and friends, and they did everything they could, ultimately the only way I could get the world to look solid again, to get myself to feel solid again, was to rely on my spiritual self—that inner airport, which was the only thing to orient by, the only thing that held steady.

But now, it seems, social media is forcing those concentric rings of the self to lose their resilient architecture, the social and material selves dominating, the spiritual self at the center like a pupil contracting to a point. A pupil that will no longer dilate, even in the darkness, even when you need it to hunt for the light.

I DECIDED ON MY COLLEGE thesis topic a few months before my eye accident, but senior fall, when my reading began in earnest, the choice felt uncanny. I was writing about how Ralph Ellison uses blues tropes in *Invisible Man*, how the underground protagonist sings himself into existence by telling his story. But passage after passage was about vision—about how others see or fail to see the invisible man, and how he eventually comes to see himself. One night as I was reading in bed, this passage in the prologue jumped out at me:

> That invisibility to which I refer occurs because of a peculiar disposition of the eyes of those with whom I come in contact. A matter of the construction of their *inner* eyes, those eyes with which they look through their physical eyes upon reality.

The narrator, a black man, goes on to relate how one dark night he bumped into a white man on the street who responded with a racial epithet, so the narrator demanded an apology, which only led to the white man cursing him, so he head-butted, then kicked the man, which led to still more cursing, so he pulled out his knife and was on the verge of slitting the white man's throat, when "it occurred to me that the man had not *seen* me, actually; that he, as far as he knew, was in the midst of a walking nightmare!"

What jumped out at me wasn't just the invisible man's recognition of the blindness of the white man's inner eyes, of how that white man saw only through a social self—with categories of black and white—which meant he couldn't really see reality at all. What jumped out at me was that the protagonist didn't get trapped by his own social self, didn't get caught in the reflection, or lack of reflection, the white man was giving him. The reflection the invisible man caught was of the trap, of his own knee-jerk impulse to respond with violence, and of how looking only through a social self means you cannot really see yourself, other people, or what is happening between you.

I wondered then, reading comfortably on my comforter in my dorm room, what it would take for me to recognize the blind spots of my own inner eye, even in dangerous situations, even if the blind spots of others threatened me. I wondered if there was a way to live from something deeper than my social self, to go through a day feeling more like I felt when I was reading, feeling that quiet of truth being revealed, regardless of where it placed me personally.

A FEW YEARS AGO, driving a scenic backroad in Virginia, I stopped at a BBQ joint not far from the Blue Ridge Mountains. There was a small dusty parking lot in front of a modest house, and a large rusting smoker not far from the door. Inside there were two tables with red-and-white-checked wax tablecloths, a tall Coca-Cola cooler with a TV on top, and a man behind the counter in a white apron.

It was only after I'd ordered and sat down by the window that I noticed the man sitting at the other table, his back against the wall. He was watching the TV. I wasn't sure he'd even registered my presence, until a commercial came on and he started to explain the plot of the show, something about the two cowboys trying to save a Shoshone squaw who was pregnant and ill. He was in the midst of detailing the backstory when the show, *Bonanza*, came back on, and he sat against the wall, attention riveted again. He was a light-skinned black man, thin and balding, probably in his fifties. The show was full of old hokey moralism and racial stereotypes. There was a Chinese cook in a little hat, and there didn't seem to be a black person for hundreds of miles.

After tucking into my food, I stole a few glances at the man to see if he was watching ironically, or putting me on. But the man behind the counter, when I brought him my plate, shook his head and said, "Every day he comes to watch. You try to change that channel and you fixin' to start a war."

"You don't mind?"

"Show grows on you," he said.

Maybe it was just the barbecued ribs and the cold Coke, but when I stepped back outside, the two-lane road looked a little wider, the blue mountains in the distance less remote. The show had grown on me too. Not because it seemed any less hokey or dated, and not because I understood why the man was so taken, why the characters' dire circumstances riveted him as they did, but because I'd caught a little of the clarity of his attention, which had nothing to do, apparently, with what he thought he was supposed to like.

About a week later, I was on the Mass. Pike on my way back into Boston, listening to WEEI sports radio. The hosts were talking about the Super Bowl. One asked Steve DeOssie, a hulking former lineman who could have played "Hoss" on *Bonanza*, about his greatest memory from the Super Bowl in 1990. DeOssie said you remember the game, sure, and there's the celebration with the guys in the locker room, but his favorite memory was after the game, wandering into a VIP room at the hotel and hearing Lionel Richie at a piano singing "Easy Like Sunday Morning."

DeOssie's voice got hoarse with feeling. He talked about how few people were in the room, he talked about the difference between hearing the song on the radio and hearing it up close, and then he gave up talking because what he was trying to express he couldn't explain. His cohosts, who usually ribbed each other for any trace of softness, any trace of not being macho and invulnerable, grew quiet. It must have been growing on them too.

WHAT I'M TALKING ABOUT isn't just being able to have your own tastes, or being able to share them without fear of mockery. I'm talking about a kind of receptivity, an openness, an ability—and it is an ability—to allow the world to reach in and nourish some central part of you, a part of you that isn't a choice.

If I could share one poem with those two boys, it would be "Gift," by Czeslaw Milosz. He wrote it late in life, living in exile in the hills above San Francisco Bay, after having been a young man during World War II in Warsaw.

A day so happy.
Fog lifted early, I worked in the garden.
Hummingbirds were stopping over honeysuckle flowers.
There was no thing on earth I wanted to possess.
I knew no one worth my envying him.
Whatever evil I had suffered, I forgot.
To think that once I was the same man did not embarrass me.
In my body I felt no pain.
When straightening up, I saw the blue sea and sails.

What starts all this is just his receptivity to the morning, to the hummingbirds by the honeysuckle—and his daily world becomes a door into his inner life, which in turn becomes a door into a larger daily world. As the speaker travels down into himself, his awareness expands: from the garden and the hummingbirds out in all directions to no thing on earth he wants to possess (past the material self), then outwards in time to forgiving everyone who has done him wrong (past the social self), then having forgiven himself through the full span of his past (and having reached his spiritual self), he moves back up into his body, feeling no pain, and now sees not just the garden but into the distance of the bay, where he notices the almost abstract forms of the blue sea and sails.

Imagine this old man now shuffling down to the local store for milk. Imagine him noticing the headlines, making conversation with the woman behind the counter. How could some of the grace from his garden not trail after him? What could be a more enormous way to inhabit a life?

THREATS

My graduating high school class was forty-eight boys, which meant we each got a full page in the yearbook. The centerpiece of my page, beside a senior portrait of me in a checked blazer, chin propped pensively on fist, was the letter that Brian Johnson, the nerd, wrote to Mr. Vernon at the end of *The Breakfast Club*.

> Dear Mr. Vernon:
>
> We accept the fact that we had to sacrifice a whole Saturday in detention for whatever it is we did wrong. But we think you're crazy for making us write an essay telling you who we think we are. You see us as you want to see us: in the simplest terms, in the most convenient definitions. But what we found out is that each one of us is a brain, and an athlete, a basket case, a princess, and a criminal. Does that answer your question?
>
> Sincerely yours,
> The Breakfast Club

The movie is a slow reveal. Five high school types we immediately recognize as they drive up for a day of detention—the princess complains in her daddy's BMW; the nerd's mother tells him he better find a way to study, mister; the athlete's dad says when he screwed around he didn't get caught, sport; the criminal shows up on foot alone; the basket case gets out of the backseat of a Caddy, which pulls away just as she tries to say goodbye—fill into five complicated vulnerable teenagers during their afternoon together in the library. Watching it, I hadn't wanted to identify with the nerd. I'd wanted to identify with the athlete, but the closest I'd come to taping a boy's buns together was being locked in a locker by the captain of the tennis team as an eighth grader. I didn't identify with the criminal or princess or basket case either, but I still wanted to identify with someone. Then Brian Johnson, the nerd, was writing that letter at the end of the movie, kissing the paper because he thought it was such a masterpiece, and I thought it was a masterpiece too. I loved the idea that high school was too soon to be able to tell a teacher who you think you are, and I loved the idea that reducing yourselves to types in the interim, to "the most convenient definitions," prevents you from discovering all the selves you might be. I could be the nerd because even the nerd wasn't just a nerd!

But something I didn't notice in high school was how much the movie depends on boredom. The five teenagers have eight hours together with nothing to do. They're bored before they even get there. They stew in their boredom. The library fills with it like a gas. They start to notice themselves—the criminal lighting his shoe on fire, the jock playing with the drawstring in his hoodie, the nerd playing with himself—and they start to notice each other. They begin with attacks along standard high school party lines, the jock even threatening the criminal with invisibility, as though he's been reading up on James's fiendish punishment:

> You know, Bender, you don't even count. You know, if you disappeared forever, it wouldn't make any difference. You may as well not even exist at this school.

But eventually boredom sputters into curiosity. They ask each other questions, and one by one, they break down, tell vulnerable stories about why they got sent to detention, get attacked some more, and ultimately get accepted. Wandering in a kind of magical high school woods—between school days, between classrooms, between cliques—they stumble upon their in-between selves. It's a high schooler's true dream of identity: not of fitting into a recognizable group, but of not needing to fit into a recognizable group, of being comfortable and accepted without a label.

Of all the parts of the movie that are fantasy, the boredom was not. We had plenty of it in the '80s. It waited between here and there. During carpool, on long bus rides to away games. It felt to me like a dimly lit corridor with a hidden exit, a door whose location I could never remember, which I eventually bumped into by taking an interest in something, even if it was something I didn't want to take an interest in, like a bird out the window or a classmate I thought I wasn't supposed to like. Then I was out, the world a little larger.

But now boredom has become terrifying to teens, itself a sign of social invisibility. If you find yourself in it, you must be unpopular, you must be doing something wrong. The shortcut out is to pick up your phone. So those doors into the unexpected, into the mysteries of the person next to you and into your own mysteries, are harder to find. Meanwhile, Mr. Vernon's prompt is always blinking on social media, everyone inadvertently pressuring each other to turn their lives into a letter they're not yet ready to write.

IN HIS NOVEL *Invisible Cities*, Italo Calvino describes a town where everyone has little choice but to live in constant awareness of their reflections. The city of Valdrada is built on the shores of a lake, and the lake's surface reflects every last detail of the city above it, not just the facades rising above the water but every room, hallway, closet. Calvino explains the effect on its citizens:

> Valdrada's inhabitants know that each of their actions is, at once, that action and its mirror-image, which possesses the special dignity of images, and this awareness prevents them from succumbing for a single moment to chance and forgetfulness. Even when lovers twist their naked bodies, skin against skin, seeking the position that will give one the most pleasure in the other, even when murderers plunge the knife into the black veins of the neck . . . it is not so much their copulating or murdering that matters as the copulating or murdering of the images, limpid and cold in the mirror.
>
> At times the mirror increases a thing's value, at times denies it. Not everything that seems valuable above the mirror maintains its force when mirrored. The twin cities are not equal, because nothing that exists or happens in Valdrada is symmetrical: every face and gesture is answered, from the mirror, by a face and gesture inverted, point by point. The two Valdradas live for each other, their eyes interlocked; but there is no love between them.

Calvino dreamed up Valdrada in 1972. Like the fifty-four other cities in the book, it's an investigation by way of caricature, by having a way of living embodied in the very layout of a city. Perhaps Calvino had observed tourists in his hometown San Remo, on the Italian Riviera, designing their experiences around the photos they could snap, or the teenagers in the local piazza preening for each other. Either way, Valdrada's lake is a prescient description of the dynamic between life off-line and life online. "Not everything that seems valuable above the mirror maintains its force when mirrored." This is where the real stakes are. Not just because we can't do justice to certain moments with online posts, but because we might stop valuing those experiences precisely because of their unshareability, leaving us to design our lives to fit into sharable frames.

Meanwhile quietly drowned in Valdrada's lake are the unbidden moments—that man seeing something in *Bonanza*, Steve DeOssie listening to Lionel Richie, Czeslaw Milosz time-traveling in his garden—and out of the lake rises a parade of moments to shape towards the identity we want to display.

To switch metaphors, every letter from the unknown parts of ourselves gets intercepted, and before we can read what is written there, we rewrite the page with our personal ad copy to send out to the world, to compete with all the other letters written the same way.

WHAT YOU CAN DO

At dinner a few weeks ago, my friend Hannah asked me, "But how do you know?" It was a breezy August night in Chicago, we were at an outdoor table, the spidery leaves from a hanging plant occasionally drifting over her forehead like bangs.

"How do I know what?"

"How do you know if an experience has touched you because it's really touching something deep inside you, or if it's touched you because of some story you can tell about it, even if that story is just to yourself?"

She's an art history professor. I didn't know if she was thinking about what moved her when looking at a painting, or about how a critic's love for a painting can get conflated with her love for how well the painting fits her theory, or if she was thinking about something else. But what sprang to mind was something I'd seen a few days earlier. I'd just come out of my front door when two girls rounded the corner on bicycles. The older girl, pink streamers on her handlebars, called to the younger, "Sabrina, sidewalk!" and the younger girl, weaving a little, cried, "Why the heck do you get road?" It was a moment that made me smile. It seemed part of no narrative about me, and was nothing I'd tell a friend, just a small moment of joy at the particularities of childhood and language, like overhearing a Norman Rockwell painting. But as I was telling Hannah about it, I realized maybe part of why I liked the moment was because Bob Coles, the professor I'd taught for, had told me about the afternoons he'd spent doing house calls with the poet and physician William Carlos Williams, and about how Williams had an ear dedicated to the homespun phrases of the patients he visited, and although none of that had occurred to me when hearing the girls on their bikes, now, as I tried to explain why the moment seemed utterly safe from any kind of self-aggrandizing personal narrative, I realized perhaps that on a subconscious level I'd liked the idea of following in the tradition of Williams, of being a listener with that kind of ear.

I told her this, then said, "But maybe it doesn't matter. If a moment reaches you, who cares what associations may have helped it reach you?"

"And if it hasn't really reached you, or only reached your ego, you know."

"I think so. Most of the time."

"And if you posted about those girls on Facebook?"

I laughed out loud.

"Why not?"

The more we talked, the more I realized what a good question it was. Not because I had a secret yearning to post, but because if I did, a post about a moment like that would almost certainly change how I thought about it. Not necessarily as it was happening, but soon afterwards, and then in my memory—a private moment turned to face outwards, retrofitted to cast me in a flattering light, those girls and their bikes trotted out as a performance, rather than allowed to gestate with other small private moments, which might increase me in some unexpected way, and which might eventually give me something more significant than self-flattery to share.

Hannah raised her glass. "To the moments we don't share with anyone."

"To the moments we don't share with anyone."

Blasphemy, we knew, in our culture of shame and sharing.

AND YET.

I could hardly tell those two boys from my high school to hoard moments, to sit on them like golden eggs that would never hatch. That was one of the problems I'd had in high school—feeling like the self I presented to classmates and to my parents had nothing to do with who I was when I read Joyce, or shot baskets in the backyard, or mulled over the day in bed. That divide, which eventually deepened into a chasm after my eye accident, was a big part of why I went to the woods. In the woods I didn't have to be seen. In the woods I could have the illusion, by ignoring my social and material self, of feeling all the parts of myself held together.

But even the spiritual self gets lonely. I didn't mind waking alone every morning, sometimes to the sight of my own breath, the fire burnt down to embers, or eating dinner alone every night. But when the day or night bumped into something new, I had no one to share it with. I remember one night, unable to sleep because of the moonlight, going up onto the roof, pushing open the screen door through the snow. There was the smell of woodsmoke, and the snow glittering blue in the trees along the hillside. Even with the brightness of the moon there was a wash of stars, a few constellations I could name and so many I couldn't—a luminous sweep of time like relatives lining up to pinch Earth's cheek, saying remember when, remember when, a luminous family, which, standing there in my snow pants on the tar-paper roof, I knew I was a part of, my tiny existence a part of something so grand. But there was no one beside me, no one to turn to, no one in whose existence, in this strange, strange, and wonderful speck of time we inhabited, to share.

CHAPTER 6

Conversational Boxes

ENDANGERED TRAITS

Large conversational boxes (Large from Latin, *largus*, large; Conversational from Latin, *conversare*, to live, keep company with, from *convertere*, to turn around; and Boxes, from Old English)

Empathy (from Greek, *em*, in, +*pathos*, suffering, experience, emotion)

BACKGROUND

If you don't speak for several days, or maybe a week, the hinge of your jaw becomes less inclined to open. I don't mean that silence breeds silence, though perhaps it does, but that physically it becomes harder to open your mouth. I'm picturing winter now, the woodstove throwing heat at my back, a blue candle on the table guttering with the draft, the windows to the trees softening with the grainy light of dusk. No sound but the popping of the logs in the stove, the refrigerator humming on, humming off. Trying to open wide and not burn my mouth on the noodles, I'd notice an ache in the back of my jaw, a slowness, like the stiffness in my fingers after coming in from the cold. It was that physical sensation that would tell me days had passed without a word.

Walks in the woods, nights by the woodstove—they stopped slipping into the shapes of language. A thought was more often an image—the sidewalk buckling with roots on my walk to school as a boy, the apple tree in the nearby meadow, its fullness of leaves and apples before the snows; the thinking just hovered around the image, perched on it and pecked a bit—but mostly there was a feeling, an atmosphere, a sense, say, of time passing or of gratitude for where I was. Every few weeks, when I did talk to a friend on the phone, the stories from his life would run as a movie in my head, and the words in my own mouth would come slowly, as though I was touching the shape of an apple tree or a sidewalk for the first time, as though I was a novice artist figuring out how to articulate the lines of the pictures I was seeing. I'd sit looking out the long windows to the woods, knowing any prospect of containing my experience was impossible. But I felt little need to try—my solitude, mercifully, felt both incommunicable and complete. My descriptions could relate a few details, and perhaps the scent of the snow and woodsmoke would carry.

But breezes don't fare well at dinner parties. When I moved back to Boston and ran into old friends kind enough to include a former hermit in their social circles, conversation didn't come easily. Small talk was a buzzing four-lane highway—career, love life, career, love life. Everyone, it seemed, was in their late twenties. I didn't know what to attribute it to—the lingo of burgeoning professional lives, post-9/11 anxiety, or my own language jet lag—but nearly everyone seemed to want the bottom line. Without the comfort of those windows to the woods, I wanted to share more than I had over the phone—to carry some of what I'd felt on the roof in Vermont into the conversation. But I had no takeaway to offer. Maybe something about how I couldn't squeeze the stars into a sentence or condense experience into a bottom line. But that answer wasn't likely to fit the size of the conversational box being offered, not over baby carrots and Chardonnay.

I hadn't thought about conversational boxes since high school. Then they were just the boxes on quizzes and tests, the blank rectangles prescribing how long your answers should be and the proportion of dread you should feel at the information you'd forgotten. But now I could feel how every question in conversation implied those boxes, and how you'd be marked down socially if you went too long or too short. And I felt as though I couldn't fit anything I had to say, and all the things I couldn't say, all the things that could carry only with the breeze, into the boxes being offered. No question my conversation skills had rusted over, adapted as they were for silence and the wide open spaces of long rare telephone calls. But I also couldn't help feeling that conversational boxes had gotten smaller. I wondered if my friends felt the same way.

A FEW YEARS LATER, texting and Twitter popularized the small conversational box. By 2007 Americans texted more often than they talked on the phone. Twitter, after being showcased at SXSW, took off. Brevity and speed of reply became cultural norms. Even NPR launched a show in 2008 called *The Takeaway*. Today the average American "consumes" about a hundred thousand words a day. It would be easy to say that the 160-character limit of SMS and original 140-character limit of Twitter inadvertently became a second kind of character limit, small boxes preventing users from sharing complex thoughts and feelings, from sharing the full range of their own characters.

But not so fast. As any poetry student will tell you, constraints often enhance creativity: Shakespeare and the sonnet. Dante and terza rima. Basho and the haiku. These poets didn't just manage within the constraints of these forms; the constraints loosed them into unlikely discoveries, acting as an impediment to certain creative impulses while forcing the development of others, just as a block to one of the senses makes the others more acute. The form wasn't a cookie cutter, imposing its little star-order on their poems from the outside, but a unit internalized by the poet as a kind of organic form, into which his mind could push and open into greater complexity than if the form hadn't existed. Creativity for these poets wasn't thinking "outside the box" but using the shape of the box to find truth through it, and letting those truths proliferate in box after box until something new, both constrained and unconstrained, appeared. "Space," as Osip Mandelstam wrote of Dante's Divine Comedy, "virtually emerges out of itself."

Of course, Shakespeare, Dante, and Basho are three of the greatest poets of all time. Perhaps more of a direct parallel to everyday texting occurred in the flourishing court culture of Japan's Heian Era (794–1185). Nearly everyone in the court wrote the five-line tanka—a poem of 31 syllables (which in English usually comes out just shy of 160 characters)—to commemorate every occasion, public and private. No part of life was complete without one, especially courtship. In her introduction to *The Ink Dark Moon*, a book of poems translated from the Heian writers Ono no Komachi and Izumi Shikibu, Jane Hirshfield describes a progression that sounds rather familiar if you replace "poem" with "text":

> The first intimation of a new romance for a woman of the court was the arrival at her door of a messenger bearing a five-line poem in an unfamiliar hand. If the woman found the poem sufficiently intriguing . . . her encouraging reply—itself in the form of a poem—would set in motion a clandestine, late-night visit from her suitor. . . . A morning-after poem had to be written and sent off by means of an ever-present messenger page, who would return with the woman's reply. Only after this exchange had been completed could the night's success be fully judged by whether the poems were equally ardent and accomplished, referring in image and nuance to the themes of the night just passed. Subsequent visits were

made on the same clandestine basis and under the same circumstances, until the relationship was either made official . . . or ended.

But this is where tanka and texting diverge. Heian court culture was highly sophisticated: Komachi and Shikibu, both women, were the most revered poets of the age, women in the court were regarded as "aesthetic equals by the men," and court members treasured the tanka poems for more than just their pragmatic or sentimental value. Like all poets living in this deeply religious society, Shikibu and Komachi wrote their poetry within a larger Buddhist context. They didn't separate the Buddhist philosophy of impermanence, which was really a way of being and of understanding being, from their ruminations on love. One of my favorite poems from Shikibu:

If he whom I wait for
Should come now, what will I do?
This morning the snow-covered garden
Is so beautiful without a trace of footprints.

You get an intimate sense of a particular woman waiting for her lover, luxuriating in the odd calm before scheduled passion, aware of her own desire but also of a kind of formless beauty, the snow-covered garden without a trace of footprints. Her attention isn't on herself or on her lover but on the physical world, and her awareness of her surroundings deepens her ambivalence about what is to come—the likely pleasure, the likely complications—until she's seeing how she's feeling in what she's seeing, all within the context of impermanence.

While the larger context for texts and tweets isn't a religion, the cult of Silicon Valley isn't far off. It has its icons, its high priests and rare priestesses, and we don't need commandments to know its values, or to appreciate the penalties for not adhering to them. The stated Holy of Holies is Speed, which takes mortal form as Speed of Writing, Speed of Reading (which for the truly righteous becomes Speed of Scrolling), and Speed of Reply. The daily catechism is roughly this:

Why is Speed good?
To share and to take in more information.
And why is sharing and taking in more information good?
To build community and bring the world closer together.
Amen.

The modified catechism, if in a private conference room in Silicon Valley, with or without ping-pong table:

Why is Speed good?
To share and to take in more information.
And why is sharing and taking in more information good?
To collect more data, sell more advertising, and make more profit.
Amen.

A Twitter company blog post explaining the company's 2017 decision to move to a 280-character limit seamlessly upholds this doctrine. There are reminders of the importance of speed and sharing, and a glorious opening adverb for a company barely ten years old:

> Historically, 9% of Tweets in English hit the character limit. This reflects the challenge of fitting a thought into a Tweet, often resulting in lots of time spent editing and even at times abandoning Tweets before sending. With the expanded character count, this problem was massively reduced—that number dropped to only 1% of Tweets running up against the limit. Since we saw Tweets hit the character limit less often, we believe people spent less time editing their Tweets in the

> composer. This shows that more space makes it easier for people to fit thoughts in a Tweet, so they could say what they want to say, and send Tweets faster than before.

The reasoning doesn't sound all that bad. But imagine a leader in the Heian Era advocating that the tanka form be made longer so that court members would spend less time editing, so they could write and send tankas faster, all of which would help them "say what they want to say." Court members would have been indignant—the constraints, and the slowness required to meet them, are precisely what enabled them to say what they wanted to say. Without those constraints, no breeze could have carried the stillness of Shikibu's snow-covered garden, not for a night, and certainly not for over one thousand years.

That speed of writing and sending tweets was the ultimate virtue for Twitter was never in question. Not for the company, and not for the app's users, apart from the Twitter traditionalists who had come to believe in the 140-limit as a kind of divine law.

Constraints on conversational boxes don't necessarily impede or enhance creativity and expression. What they do is concentrate the cultural values behind those constraints.

A FEW YEARS AGO, after a night of gusts and whipsawing rain, the pedestrian sign around the corner from my apartment hung upside down. But like a clever vandal, the storm had perfectly aligned the sign on the pole. Instead of walking safely across a solid black line, the figure now strolled upside down along a minimalist sky. On the far side of the street, Starbucks, on the near, the Levine Chapels funeral home. The crosswalk suddenly looked designed for Jewish souls on their way to grab a latte. I noticed this on my morning walk. Passersby wove around me, eyes on cellphones or fixed straight ahead, their routines apparently connecting them elsewhere. Some stepped without hesitation into the crosswalk, joining whatever invisible souls were already there. My inclination was to share a look with someone on the street, but no one's eyes were available.

What if I snapped a photo, I thought, and posted it on Twitter? The attention I would get for my attention! But how would the photo read? I knew whatever you tweeted had to be quickly identifiable. The labels came pre-made in a way, ready to be affixed to the small boxes. Some tweets, like those of President Trump, tried all of the apparent options at once: further evidence, humor, grievance, call for support, and (wounded) pride. Could I post something that wouldn't get stuck with any of them?

Meanwhile people kept stepping into the crosswalk, unaware of the imaginary souls in their company. And I liked watching people on the sidewalk blithely pass the sign, as though upside-down soul walking were as much a part of our world as the park, which was signaled by the next sign, a seesaw. Maybe I could borrow a friend's phone. Maybe what I needed was video and a GIF. Maybe a tanka.

After the spring night storm,
the morning rush. A pedestrian sign
hangs upside down.
Between Starbucks and the mortuary
No one minds.

I imagined it coming up on a friend's Twitter feed—in a small box between small boxes of Trump outrage and quips about the new hipster bar. How would it read? Probably as a failed joke. Or as a warped kind of metaphysical pride. Or, most likely, as further evidence of someone who didn't know how to use Twitter.

WHY IT MATTERS

Imagine you're a tourist in a fantastical city. Every morning as the fog burns off you notice the signs have reversed directions from the day before. Not just every pedestrian sign, street sign, and restaurant sign, but every unwritten sign—every unspoken trust, every tacit assumption, every unwritten rule. The powers behind the overnight changes are sometimes visible but more often not, the shifting factions within each group even harder to grasp than the city's protean map.

It would be comforting to call this city Adolescence.

It would be more apt to call it the US circa 2020.

Now imagine a magical guidebook. It changes as fast as the city does—with insights, recommendations, analysis. For every new restaurant, rule, or upheaval, you can read the hot takes of your favorite experts, friends, or enemies. Guidebook in hand, you can sit at a busy café, or anywhere else, and wrest a semblance of order from the throbbing confusion. This is what makes the guidebook truly magical. You see your very own personalized city. Based on the parts of the guidebook you've enjoyed most (and where you've spent money, and lots of other little magic things no one really understands), the guidebook instantly removes the unsightly from your sight—why bother with an opinion that upsets you! If you fear the guidebook sounds limiting, with nothing to define yourself against, fear not—the best in personalization means having the best in enemies, the easiest to demonize, who scheme in ghoulish outline in the distance. The voices that agree with you become clear and distinct; the voices that disagree with you, a foggy blur. Even when excursions into the unpredictable streets of the actual city, with chance run-ins with unwieldy people who don't fully match up with the guide, prove surprisingly pleasant, the evening ends with you back in bed curled around your guide, grateful for its familiarity and comfort, its sense of containing your city, the one that nobody else can see.

THE SUMMER AFTER HIGH SCHOOL I was a counselor at an overnight camp in New Hampshire. The kids played sports on fields surrounded by trees—to hit a homerun was "to woods" it—swam in a lake so cold it turned Johnny Bent's lips blue as a raspberry popsicle, and learned to square dance in an old wooden hall that effectively became a sauna on Saturday nights, the line for the water fountain snaking into the mossy darkness at the edge of the trees. The boys in my cabin were twelve, edging with fear and bravado into puberty, and at night the little wild men in them came out. Anthony liked to stand atop his cubby naked, shake his hips, and proclaim himself King of the Jungle! Kevin and Josh liked to howl like coyotes. The only way to get them to brush their teeth and into bed was to threaten not to do cabin chat. That this transformed the wild rumpus back into a bunch of boys putting on their pajamas amazed me.

Cabin chat just meant lights out, a candle on the floor, and a conversation. One night I played Springsteen's song "Walk like a Man" and asked when they thought adulthood began ("When you're married!" "When you're eighteen!" "When you have responsibility!"). Another night, the question was about inner riches (like love, contentment, kindness) versus outward riches (like money, cars, houses): if a genie appeared and promised you a fixed total of riches, what proportions would you choose? (Most boys hedged towards a fifty-fifty split, but some extremists, like Anthony, went for mansions, Ferraris, and hot tubs, pointing out that these would bring you love and happiness, duh.)

The boys would rush into bed for these chats. In the dark, under their covers, they grew remarkably candid and vulnerable. Boys who peacocked around the basketball court during the day shared fears of their parents' divorce or of not having close friends when they went back to school. They talked about things they would change about themselves if they could, even Anthony. They weren't anonymous in the dark—their voices and the locations of their beds were all recognizable to each other—but they were less exposed than during the day.

If Mark Zuckerberg wandered into Bunk 8, and slid into a sleeping bag beside the candle on the floor still in his designer jeans, he'd likely say Facebook works the same way. You're not anonymous, but you're less exposed, and you can share whatever you want to share, and your "friends," who are safe in their own bunk beds or on their private jets or wherever they are, can respond with encouragement.

But maybe Anthony would point out that while the private jet sounded awesome and he was planning on buying one or two himself, definitely with lots of champagne and a hot tub, there was kind of a difference. With your friends in the cabin, you're going to see them tomorrow, and the next day, and for the rest of the summer, so you want to get to know them because you're, like, a cabin, and Bunk 8 rules! And maybe Kevin would point out that the jet's really cool, and could you really have a hot tub on a plane?, but one of the things he likes about cabin chat is the darkness, and looking out the window above his bed and seeing the stars coming out above the trees, and how you kind of need your imagination to know the other guys, to think about all the parts of their lives you don't really see, like who they are back home and when they're going to school, and how you couldn't really just combine cabin chat with Camp Meeting, where we get all the news reports of the week, who hiked what mountain and who beat who in the basketball challenge, and still feel that mystery about everyone as they talked. And maybe Josh would ask if conversations on Facebook were really totally private anyway—I mean, like, how did Mr. Zuckerberg even know to come into our bunk?

NUMEROUS STUDIES, seventy-two to be exact, suggest that empathy in college students has dropped 40 percent in the last thirty years, with the steepest drop-off since 2007, which also happens to be when the iPhone hit the market. That our boxes for each other are getting smaller is hardly Zuckerberg's fault alone. MIT psychologist Sherry Turkle cites a few commonsense reasons for our empathy plunge:

> The presence of a phone, even if off and upside down on the table, changes how we talk to each other. Sensing interruption could come at any time, "we keep conversations light, on topics of little controversy or consequence."
>
> The more options we have for conversations that don't happen face to face, the more we don't have conversations, especially difficult ones, face to face. Texting feels safer.
>
> In an effort towards empathy, towards saying the "right thing," we don't confront tricky topics in person. Learning how to sit with lulls and awkwardness, with not knowing what to say, takes a hit, and, in the long run, so does empathy.

One thing Turkle doesn't mention is that measuring empathy is an uncertain business. You can't stick a thermometer in someone's feelings. One of the primary empathy studies, for example, repeated year after year, simply asked students to rank how well descriptions, such as those given below, described them.

> I often have tender, concerned feelings for people less fortunate than me.
>
> I would describe myself as a pretty soft-hearted person.
>
> When I see someone being treated unfairly, I sometimes don't feel very much pity for them.

What's fascinating is that students in the study could easily have presented themselves as more empathetic if they'd wanted to. They just needed to mark those descriptions as describing them extremely well. But at least the more contemporary students didn't do that. Maybe what's really changed is honesty in self-reporting—earlier generations artificially boosting their empathy scores, and today's college students disclosing the full frostiness of the human heart. But I doubt it. Why would the level of honesty change so dramatically? But the boosting direction, determined by student values, might have changed, at least a bit. My bet is the 40 percent empathy gap is the combination of a drop in empathy along with a drop in how much college students value empathy. With more pressure on success, greater income inequality, and greater emphasis on efficiency, it's easy to imagine an eager freshman, shirt artfully half-tucked, hurrying into the study from his Computer Sci lecture, thinking: *better not to say I'm pretty soft-hearted, better not to waste time on tender feelings for the less fortunate.*

The ghost of Frederick Winslow Taylor, stopwatch dangling from his hand, has finally crept into our interactions with each other. Empathy is inefficient and slow. Empathy is unwieldy and unpredictable. It doesn't fit in the schedule. It happens in a kitchen late at night, or on a car ride too long not to talk and not to listen, and usually involves stretches of silence, and nearly always involves the body, whether a glance, or a hand on the arm, or maybe a full snot-running hug, a return to the body as human and loving instead of violated or shameful, a return to tenderness and fragility and a nonphysical kind of strength, with its understanding that sometimes you cannot improve the silence and you do not need to. Empathy also requires humility—an awareness that you can't know everything about someone and that part of understanding is allowing for all you cannot know. To struggle to give a helpful answer, or to listen when there is no answer to give, was always somewhat uncomfortable—but it's more uncomfortable now that our brains are wired for interruption, for avoiding vulnerability, and for thinking there's a "right answer" we should give quickly.

So we say less about the hard stuff, are worse at listening to the hard stuff, and our range of empathy contracts. Meanwhile we're having more "conversations" than ever.

SOCIAL ACTIVISM MAY BE THE NEW EMPATHY—the new way of showing you care. An unheard voice becomes a chorus. Evidence mounts. Movements form. People find community, support, and strength. Between three million and five million people participated in marches around the country as part of the Women's March protests the day after Trump's inauguration. More than one million people participated in the March for Our Lives less than two months after the Parkland shootings. This kind of rapid organization and planning couldn't have happened without social media. Twitter and Facebook helped people turn to each other, find each other, and demonstrate.

But as several recent books have pointed out, these impressive numbers are a testament less to the power of the movements, and their ability to effect change, than to the power of social media itself. In *Twitter and Tear Gas: The Power and Fragility of Networked Protest*, Zeynep Tufekci writes, "In the networked era, a large, organized march or protest should not be seen as the chief outcome of previous capacity building by a movement; rather, it should be looked at as the initial moment of the movement's bursting on the scene, but only the first stage." By contrast, Tufekci points out that the Montgomery bus boycott of 1955 and the March on Washington of 1963 took civil rights workers years to build the infrastructure of organizations, leadership, and cooperation that would survive the hardships of the boycott, which lasted over a year, and the challenges of carrying out the march.

The traits those civil rights workers had to learn—effort, patience, courage, and sacrifice—are not the traits social media movements develop, yet they may be the very traits social movements need to last. On the other hand, social media plays easily to the long-term goals of repressive movements, which don't require long-term discipline or courage or sacrifice but only volume and persistence, "deliberately sowing confusion, fear, and doubt by aggressively questioning credibility, . . . creating or claiming hoaxes, or generating harassment campaigns." Tufekci wrote those prescient words before the news about Russian interference in the 2016 presidential campaign, before the revelations about bots marching unrecognized through our feeds; with social media's power of amplification and simplification, or outright misinformation, she could see the hate speech and hoaxes on the wall.

If activists are willing to use social media as a starting point rather than as their chief tool, perhaps the power of social media for organizing will outweigh its power to sow fear, its power to deepen our primal instincts towards tribalism. But that will require far more than turning off our screens sometimes; it will require spending enough time face to face with each other to develop the strong bonds, the patience and courage and willingness to sacrifice, that allowed civil rights workers to link arms, and to keep linking arms, and endure.

As for social activism as empathy, it can certainly be a start. But as with activism itself, the one hundred likes or retweets rallying to your support won't mean much unless, sooner or later, there's someone sitting beside you in a kitchen late at night or on a long drive somewhere you need to go, someone willing to listen, not to the cause, but to you.

THREATS

If the only way to reach a large audience is through a small conversational box, and the argument against small conversational boxes can't fit inside a small conversational box, then how do you fight the small box invasion? My fear is that we'll just respond to small boxes with more small boxes, like an immune system inadvertently attacking what it's trying to defend.

However disturbing on a personal level, the consequences are even more dangerous on the public level.

Here's an example. On a steamy Labor Day morning in New York City, the *New Yorker* announced its invitation to former White House chief strategist Steve Bannon to speak at its 2018 festival. Editor David Remnick planned to interview Bannon on stage in what would be "a serious and even combative conversation."

Bannon is nothing if not a small-box king: a true megaphone for white identity politics and nationalism. "Let them call you racist. Let them call you xenophobes. Let them call you nativists," he has said. "Wear it as a badge of honor. Because every day, we get stronger and they get weaker."

Could such a conversation, between an internationally renowned journalist and an internationally renowned hatemonger, be useful? Could putting tough questions to a powerful person in a public forum help hold the person to account? Or was it just a remarkably bad idea? How should the media handle leaders who spew strains of anger that are highly resistant to reason, evidence, and decency?

These conversations about conversation, sadly necessary for our times, never happened, at least not in a public forum. What did happen was a small-box outbreak as soon as the Bannon invitation was announced. Celebrities invited to the New Yorker Festival tweeted they were backing out:

"Maybe they should read their own reporting about his ideology," tweeted Judd Apatow.

"I'm out. Sorry, @NewYorker. See if Milo Yiannopoulos is free?" tweeted Patton Oswalt.

That still steamy Labor Day evening, after a day-long Twitter outcry, Remnick pulled the invitation. The small-box outbreak had become an epidemic. Outrage, sanctimony, cancel my subscription! Impressive linguistic and moral one-upmanship. Platform became Legitimization, Legitimization became Imprimatur, Imprimatur became More White People Bullshit!

Likewise, on Bannon's side: Cowards became Gutless Liberals, Gutless Liberals became Commies, Commies became Deep State!

Not that I'm suggesting a moral equivalency (one of the top Twitter sins), just a conversational pattern. We're caught in a doomed game of Scrabble: everyone gets a handful of concept tiles—not picked at random but by cultural identity and political affiliation—and it's up to us to arrange those tiles as artfully as possible at every outraged turn. If you question someone's application of a concept, e.g., "normalization of hatred," to a particular event, then you're dismissed as not understanding, or not capable of understanding, the tile itself. The more craftily you play, the more social points you get! We seem to have internalized, from our own private data analysis, what kind of response gets likes and retweets, and what gets shot down, and, more and more computer-like, we play the right move.

> The praiseworthy attempt to see things in their wider context becomes so formalized that instead of applying that technique in particular, unique ways, appropriate to a given reality, it becomes a single and widely used model of thinking with a special capacity to dissolve . . . everything particular in that reality. Thus what looks like an attempt to see something in a complex way in fact results in a complex form of blindness.

That wasn't David Remnick. It was Nobel Prize–winner Vaclav Havel on the state of public discourse in Prague in 1965. A playwright imprisoned for his political convictions, Havel spent three and half years in jail, where he did manual labor and was routinely beaten. He lost a lung and nearly died. He did not take oppression, or the normalization of hatred, or speech of any kind, lightly. Perhaps he would not have invited Bannon to speak at a festival. Perhaps he would have. Either way, he knew that evasive thinking and phraseology only strengthen oppressive forces. On the difficulty of editing a literary magazine in Prague in those years, of trying to print pieces without "concessions—regardless of which side they are intended to satisfy," Havel wrote:

> I wasn't really prepared for how difficult it has been for the literary climate to come to terms with something as natural as the attempt to speak without using established phraseological evasions; for how this climate—still dominated by evasive thinking—has suddenly begun to cling to these evasions: for how hard it is to come to terms with someone actually enjoying the luxury of openness; and for the irritation, the sour looks, the snide remarks evoked by something so down to earth as a group of people attempting to be true to themselves, without having to cut a deal with the literary climate.

Remnick may well have been thinking similar thoughts. But he couldn't say them, not in 2018 in the US, not even as the editor of one of the country's most influential magazines. He issued an apology for the Bannon invitation. The *New York Times* opinion headline followed swiftly: "Now Twitter Edits the New Yorker." But that headline, and the snide glee of the piece itself, was angling for Twitter glory too.

How can we have a meaningful conversation, one that allows for both discomfort and discovery, one with "the luxury of openness," on a Scrabble board?

IN HIS ESSAY COLLECTION *The Braindead Megaphone*, George Saunders asks us to imagine a man sitting alone in a room. Through the open window, he hears someone shouting information about the house next door. He begins to imagine the house based on what he's hearing. Saunders asks: "What are the factors that might affect the quality of his imagining? That is, what factors affect his ability to imagine the next-door house as it actually is?"

Saunders helps us out with the answers: clarity of language, lack of agenda of the guy describing the house, time and care the guy takes to construct his message, and total time allowed for the communication. Saunders helps us out some more by taking the thought experiment to its logical end:

> So the best-case scenario for acquiring a truthful picture of that house next door might go something like this: Information arrives in the form of prose written and revised over a long period of time, in the interest of finding the truth, by a disinterested person with real-world experience in the subject area. The report can be as long, dense, nuanced, and complex as is necessary to portray the complexity of the situation.
>
> The worst-case scenario might be: Information arrives in the form of prose written by a person with little or no first-hand experience in the subject area, who hasn't had much time to revise what he's written, working within narrow time constraints, in the service of an agenda that may be subtly or overtly distorting his ability to tell the truth.

Saunders wasn't making an elaborate metaphor about Twitter storms or even about Trump. He wrote the essay before Twitter existed, before the notion of Trump as president was anything more than a *Simpsons* episode. He was writing about our decision to invade Iraq in 2003.

But now two-thirds of Americans get some of their news on social media. Which means when we're not listening to a shouting man, we're struggling to hear someone above the shouting man. And our options for response are remaining silent or shouting back ourselves. As Nicholas Carr wrote during the 2016 election, in a warning that now seems eerily both too early and too late:

> When we go on Facebook, we see a cascade of messages determined by the company's News Feed algorithm, and we're provided with a set of prescribed ways to react to each message.

> We can click a Like button; we can share the message with our friends; we can add a brief comment. With the messages we see on Twitter, we're given buttons for replying, retweeting and favoriting, and any thought we express has to fit the service's tight text limits. . . .
>
> Because it simplifies and speeds up communications, the formulaic quality of social media is well suited to the banter that takes place among friends. . . . But when applied to political speech, the same constraints can be pernicious, inspiring superficiality rather than depth. Political discourse rarely benefits from templates and routines. It becomes most valuable when it involves careful deliberation, an attention to detail and subtle and open-ended critical thought—the kinds of things that social media tends to frustrate rather than promote.

Now that guy shouting next door, or worse, a series of shouting guys, a whole neighborhood of them, is our peephole to the world. And as we respond to them, it's hard to not become a shouting guy too.

WHAT YOU CAN DO

We need a tool to communicate better. That's what drove the poets who invented the tanka, the terza rima, and the sonnet; it's what drove the camp counselor, whoever he or she was, who invented cabin chat; it's what drove the camp directors who invented Camp Meeting based on the old New England town meetings; and it's what drove the Puritans who invented the old town meetings for town residents to discuss, in a civil manner, pressing town needs. In every instance, the inventors sized the communication box to fit the particular needs of the kind of communication they wanted to foster: Shakespeare changed the sonnet form to have three quatrains and couplet so he could develop more complex ideas than Petrarch; the counselor developed cabin chat (lights off, candle on the floor) so the campers would be more open than they were during the day; and the Puritans invented town meeting so residents could talk about their issues face to face, with the reminder of their moral and civic responsibility to each other present in the very building where they met, in the very reason for the meeting itself.

What baffles me is that we haven't yet turned the power of digital technology—specifically its visual appeal and complexity, its potential for displaying myriad facets of an argument and linking to supporting facts—towards apps that foster the kind of political conversation that will promote democracy rather than demagogues. A nonpartisan group of political science professors and policy experts, helped along by Silicon Valley's sharpest young people who actually care about democracy, could design an app for the 2020 elections. Maybe the homepage takes you first to the issues—you click on Economy, or Foreign Trade, or Climate—and you would find facts, useful background information, questions we're facing, policy approaches with arguments and counterarguments, all written in clear nonpolarized language, complete with argument maps, so you could see how one idea relates to another, and also with gray areas clearly marked, candid acknowledgments of what we do not know or have a hard time predicting. Then, and only then, could you click on a candidate—being reminded of the danger of single-issue voting if you've only clicked on one issue—and you could see highlighted the policies and arguments that particular candidate advocates, and where they might have contradicted themselves, and where the facts they've used in their arguments haven't been accurate. Perhaps you could also fill out a page, after reading about several issues, for the policies you'd prefer, and the app could show you which candidates hold similar positions and why, along with the counterarguments for those positions.

Such an app would bypass the cult of personality, bypass the shouting guys, and would lead to clearer thinking, which would lead to clearer conversation, which would lead to more informed voting. Classes on the app could be taught in high school, college, and continuing ed programs—it could be, amazing to say, as integral to our daily lives as Facebook and Twitter are today.

What I'm saying is that we need to make our own boxes: Boxes large enough for the political moment. Boxes that haven't been fabricated in the factory of speed and efficiency, or in the factory of attention manipulation. Boxes that have democratic values—like fact, reason, and even empathy—behind them. And, more broadly, boxes that come in a vast range of sizes for our almost infinite range of communication needs, for the complex political analysis, the whispered endearment, for the tenderness that opens beyond boxes entirely.

AT A READING at the Brookline Booksmith in 2012, Etgar Keret faced one of the Top 5 Dreaded Q&A Questions for authors.

"Where do you get your ideas?"

I'd wandered into the bookstore hoping for inspiration or at least distraction. The reading had delivered both—laughs and an undercurrent of mystery, a reminder of a hidden realm just behind our daily lives, which Keret's stories seemed specially positioned to reach. But now a well-intentioned audience member wanted the magic trick behind Keret's grown-up fairy tales explained.

Keret has the large expressive face of a character actor, and he leaned over the lectern as though to draw in more of the question. Then he said that in his daily life in Tel Aviv sometimes he'd see an interaction on the street or overhear a conversation in a café that intrigued him. That night, over dinner, he'd tell his wife the story. "And if she tells me my story is boring, then I know I've got a story to write."

The man who asked the question laughed uneasily. But Keret wasn't joking. He went on to explain that if he wasn't able to convey to his wife the seed of what had intrigued him, if he couldn't make it meaningful to her, then he probably didn't understand the moment itself—why it had caught his attention and lodged in his memory. So he was on to a kind of mystery.

I thought of one the stories he'd read, "Suddenly, a Knock on the Door." In it, three armed men, two with guns and one disguised as a delivery guy who slips a cleaver from a pizza box, demand that a character named Keret tell them a story. In the Middle East, one of the men says, brute force is the only language people understand. But the story is unsettling beyond culture or politics—perhaps there was a personal moment Keret the author had witnessed that led him to wonder about what happens when something that can't be gained by force, like the imagination or memory or love, is demanded by force. Maybe he'd overheard a father commanding his son to retell a story for friends at a café. Maybe he'd heard a young man pleading with a woman to love him. Perhaps Keret couldn't fit the full uneasiness of such a moment, its personal and political resonance for him, into dinner conversation with his wife. How could anyone fit it into dinner conversation? But he didn't give up on trying to articulate it. Given what he said at the reading, maybe it was the very boxlessness of his uneasiness that reassured him he was on to something worth saying.

This is what no one tells us anymore. The importance of the boxless, of our complex uneasiness, of the invisible cobwebs wiped hastily from the face. Those are the stories we need to tell. Those are the stories we need to hear.

SO, ONE MORE STORY OF ENCOURAGEMENT. A story I've always loved for its Houdini escape from a box that seemed utterly inescapable.

The scribes and the Pharisees bring Jesus a woman taken in adultery. They've got the violence hunger of a mob, and they've also got old-school cunning. They want to see the woman punished, but what they really want is to trap Jesus. They know the law calls for stoning a woman taken in adultery, and they know that Jesus's teachings call for mercy. Either Jesus will have to defend the woman and betray the law, or he'll have to condone the stoning and betray his teachings. Or perhaps he'll keep quiet and betray himself as not being a leader at all.

Defend, attack, or keep quiet. The scribes and Pharisees say nothing about a 140-character limit, but the constraints aren't all that dissimilar.

But Jesus does something strange. He kneels down and draws in the dust with his finger "as though he heard them not." Maybe he's stalling for time. Maybe he's giving the men in the mob, and the woman, time to reflect on their behaviors. Either way, his first response to their provocation is to pause, kneel down, and show humility. (Modern footnote: A Google search for "pause for humility" gets a little less than two million results; a Google search for "rush to anger" gets over twenty-seven million, with videos of Rush Limbaugh topping the list.)

The scribes and Pharisees continue berating Jesus—they want an answer: The law commands stoning a woman caught in the act. What is his response? Jesus stands and says, "He who is without sin among you, let him cast the first stone." Then he kneels down again. This is where the story gets really strange. His moral horizon is too large for him to get caught in the personal—not just with the woman but with the men. He doesn't condemn the woman for adultery, and he doesn't condemn the men for their bloodlust or their scheme to trap him. He doesn't preach or even criticize. But the scribes and Pharisees hear what Jesus is saying to them. And "convicted by their own conscience," they leave the scene "one by one," as individual after individual, rather than as an angry mob.

Now, I'm not naïve enough to expect us always to speak from love and mercy. Or to imagine that if we did, members of angry mobs would suddenly see themselves and their targets as individuals, and repent and change their ways. If you knelt down in front of Steve Bannon, you'd probably get kicked in the teeth.

But for gaining some measure of control over our own responses to the latest provocation, the story of the adulterous woman isn't a bad escape story. For the adulterous woman, for the angry mob, and for Jesus himself.

He made a new box—and it set him free.

CONCLUSION

Let's Make a Deal

FROM OUTER SPACE, the Atacama Desert in Chile is a parched brown scrape, a six-hundred-mile scar on our bright blue spinning planet, as though a carpet sample of Mars had been laid down billions of years ago and forgotten. From the thin air of the desert, the stars look both very close and very far: the starlight that reaches the telescopes on the mountaintop observatories arrives from thousands of years in the past. The recent past is strangely near and far too. In the mid-'70s, General Augusto Pinochet sent thousands of political prisoners to a concentration camp in the nearby abandoned salt-mining town of Chacabuco. Most of the prisoners "disappeared," their bodies buried in mass graves, but because the evidence refused to decompose in the desert, the bodies were dug up and dumped in the sea. However, remnants of bone and clothing remain. Just as astronomers search for the past in the skies, groups of elderly women, with shovels as their telescopes, search for the past in the sand.

Late in the documentary *Nostalgia for the Light*, which tells this story, we're introduced to the astronomer Valentina Rodríguez. In 1975, when she was a year old, the Pinochet police detained her grandparents, demanding to know the location of her parents' home. When they didn't answer, the police threatened to take the girl. The grandparents answered. The grandparents raised her. The parents never returned.

Rodríguez, now a mother, says in the film:

> Astronomy has somehow helped me to give another dimension to the pain, to the absence, to the loss. Sometimes when

> one is alone with that pain—and these moments are necessary—the pain becomes oppressive. I tell myself it's all part of a cycle which didn't begin and won't end with me, nor with my parents, nor with my children. I tell myself we are all a part of a current, of an energy, a recyclable matter like the stars which must die so that other stars can be born, other planets, a new life. In this context, what happened to my parents and their absence takes on another dimension. It takes on another meaning and frees me a little from this giant suffering, as I feel that nothing really comes to an end.

I saw *Nostalgia for the Light* in March of 2015, shortly after one of my closest high school friends was killed in a train crash. He was coming home from work, sitting in the first car of the Metro-North from Manhattan. I saw the news reports—the mangled train, the blackened skeleton of the SUV, the black clouds of smoke—before I knew Joe was on board. I was at the gym. A bank of large-screen TVs above the ellipticals, a dozen or so of us jogging in our own tiny orbits to another tragedy, grim-faced reporters in winter parkas at the scene, red flashing lights glaring in the smoke behind them. The next morning, after finding out Joe had been in the first car, I clicked from story to story, national to local. So much information, so few answers.

The funeral was a help. More than half of our class of forty-eight boys showed up. The firm handshakes, the solid claps on the back. We'd grown into the blazers and ties we'd been forced to wear so often as boys, into the solemnity of those Halls and Latin songs. We'd grown into what as teenagers had seemed an unfair burden—our lives' interdependence, our responsibility for each other. After the ceremony, we retreated to a local bar and stayed through the evening no matter the snowstorm pushing up the coast. The stories kept opening into other stories, as though we were participating in a group dream, keeping it aloft—until I could feel Joe's presence, feel in my body how I used to feel sitting next to him in Hall on those hard wooden seats, the blue Latin vocab booklet tucked by his thigh so he could cram, or sitting next to him in his car The Cloud, with its smell of vinyl and French fries and lust, until I could feel us on our high-speed escapes during free periods to McDonald's, Joe wearing his ridiculously large aviator sunglasses and goofy smile, until I could remember the way the school's narrow corridors looked different, conquered, as we signed back in.

The snowstorm started up outside the bar's windows, big flakes blowing sideways under the streetlight. There were no stars in the sky, but we were finding more space and more time in each other. No one stood up to leave. No one mentioned the drive back to Boston or Brooklyn or Manhattan. We talked into the night. We talked into the storm.

IT'S IMPOSSIBLE TO KNOW how space felt to people in earlier generations. How far away was far before anyone crossed the ocean, before anyone crossed a continent on a train, before anyone circumnavigated the globe in an airplane? How much closer did the stars get thanks to Apollo 17 in 1972, when we saw the first full-view photograph of Earth from space? "Once upon a time, in a land far, far away," stretched in *Star Wars* in 1977 to "A long time ago in a galaxy far, far away." Our fantasies suddenly needed more space to stir the same faraway feeling, the same mystery. Science fiction had to keep exploring alongside technology, so we could bring the cosmos closer and yet still dream just as far.

How far away is far today? My father wrote on my mother's high school yearbook page in 1962: "I love you to the moon and back." To a high school sweetheart in 2018, that would probably sound like an Elon Musk reference, like a possible vacation if they were still together a few years down the road.

How much space do our dreams and griefs need to have to have room enough? Now that we've pulled our daily maps, the world map, and even the stars so much tighter, what happens to the parts of our lives that actually benefit from space and time?

IN 1975, the same year the Pinochet police came for Valentina Rodríguez's parents, Andy Warhol was thinking about the allure of more space, the kind only technology could offer:

> If you were the star of the biggest show on television and took a walk down an average American street one night while you were on the air, and if you looked through windows and saw yourself on television in everybody's living room, taking up some of their space, can you imagine how you would feel? . . . No matter how small he is, he has all the space anyone could ever want, right there in the television box.

Of course, now the possibility of being on screens in people's homes, and everywhere else they go, is exponentially greater, has been "democratized," thanks to social media and YouTube. The shared vastness of the night sky has been pulled inside out not just into television sets but sifted into little flashes of light in everyone's pocket. We can each be a star, and the hope is to be seen by all the other stars, to be seen everywhere and at different times. That's the new extra dimension for our pain and wonder, our substitute cosmos, and it requires almost no waiting, almost no vulnerability to be told "you're not alone" as you sit alone in your room posting that you are lonely. You can have the reassurance of seeing yourself being seen.

Meanwhile the near and distant stars in the Atacama Desert go on wheeling overhead. Every night the temperature drops, and every day the temperature rises, and the salt-encrusted earth groans as the ground expands. The desert's rasping breath might be as ancient as any sound on the planet, a sound older than any form of hearing, older than any form of needing to be heard. The portals in our pockets, even when they give us images and sounds of the Atacama, can't offer that context Rodríguez talked about, the feeling of being part of a universal cycle, of being part of the immeasurable. They're too good at responding to our immediate needs, too efficient at doing our bidding, for us to be getting lost in something larger than ourselves.

IN 1963 NBC PREMIERED a game show called *Let's Make a Deal.* It began with a prerecorded segment of the host, Monty Hall, sitting alone in an auditorium with a long, thin microphone, pitching the show as a microcosm of American commerce and culture:

> This is television's only trading floor, where every day individuals who control the finances of America—the women of course—come to make deals. And what's more exciting for a woman than trading, or swapping, or looking for a bargain? It's suspense every second. . . . We'll buy, sell, or trade everything from aardvarks to zithers.

The weirdly sexist pandering to the housewife demographic is obvious; why watching people barter would be "suspense every second" is not. The show's very first trades were with a well-dressed young woman from the audience, Maggie from Rialto, California. Monty Hall offered her $5 for anything in her purse with her name on it. A bit eerie now that the first request was for personal information; stranger still, the next trade: $30 if Maggie could tell Monty how many letters were in her full name; and strangest and most exciting of all, would Maggie trade the $35 now held in her stunned hand for what was inside a large unmarked box?

"I'm a gambler!" Maggie said.

And for her gamble, Maggie would take home . . . A NEW FUR COAT!

Let's Make a Deal would eventually spread to homes in twenty-one countries. A remake in the US began running in 2009 and is still thriving. Audience members traded in their business attire for zany costumes. The new host traded in Monty's lounge lizard jokes for charm and improvisation. But the still-beating heart of the show was right there in that first exchange: the thrill of trading in unknowns. Never is a contestant asked if she wants to trade a toaster for a baseball glove, a chocolate cake for a picture frame. She's asked if she wants to trade the $200 in her hand for what's inside an unmarked box, or, harder still, if she wants to trade what's inside the unmarked box for what's behind curtain number 3.

The show isn't a model of commerce unless you shop in stores with concealed merchandise; it's a model of a particular kind of decision, the kind we have to make in every long-term prospect of our lives—in our relationships, in our careers, and in our technology—when we can't know exactly what we'll be giving up or getting in the bargain. The show reveals what's really inside box number 1, and what's really behind curtain number 2, and what's really behind the biggest curtain of all: time. Maybe you get a Brand New Car! Maybe you get a hamster running on a wheel. Either way, you know instantly whether or not you made the right call—a final calculation of consequences rarely available in real life, certainly not instantly, and sometimes not at all.

Now a new Monty is asking us what kinds of trades we're willing to make on our very orientation in the world—on our attention, on our sense of time and space, on our sense of identity, on how we know what we know, and how we talk about what we talk about. He's put a small shiny new box in our hands. It feels marvelous. Don't we want to make the trade? Aren't we gamblers? We'll just have to give up a small fraction of whatever's in the cumbersome box of our brains and allow it to be rewired. How about it? Aren't we gamblers like Maggie?

How to decide when we can't really see both sides of the trade? When the long-term consequences are impossible to predict, no matter what the new Montys promise? What should we steer by in these decisions? Where are the stars for the stars in our hands?

I PROBABLY MADE the right decision in returning from the woods, but there's no fur coat or brand new car to let me know for sure. My morning walks along the community golf course are shorter and less wondrous than my excursions through the uninhabited Vermont hillside. I see more dogs, fewer deer. My mind is frequently less clear, less open, and, strange to admit, less loving than it was in the woods. But I see things here sometimes, small moments that remind me of all I've returned to.

A few days ago, one of the last golden days of autumn, I saw a girl in the park standing strangely still, her back to a massive oak tree. She was maybe five or six years old, in a sundress, and she stood looking down at the roots, or maybe at her feet. There were no other children in the park. Sun, a slight breeze, a faint stirring in the canopy of leaves. As I came down the sidewalk, I could tell she wasn't really looking at anything. She had a very quiet smile on her face, like she had the most beautiful secret. As I walked past her tree, she didn't move at all. Then I heard a voice counting, and near the swing set, hand over his eyes, I saw a man who was probably her father.

Knowing he would come looking for her, knowing she would be found, the girl could revel in her solitude. Her smile was so inward, so peaceful. As I sat down with my lunch, I imagined her as an adult, still having that faith, still trusting that those who loved her would be able to find her, or that she would be able to find them, even after she'd been alone with the trees and her own thoughts. And I imagined a boy in another park, behind another tree, growing up the same way. And then thousands of boys and girls, behind thousands of trees in thousands of parks, all growing up with this trust in being found, all allowing their own thoughts and eventually their own bodies to wander and get lost, and I thought about the experiences, and all the different ways of looking at things, they'd be able to share with each other once they returned.

This is what I've found most moving about my own return from living behind the trees—all the people I've found who are trying to figure out how to live with their own kinds of solitude, whether in the city or in the woods. What I want us to protect isn't just the distinctive range of consciousness within each of us but also the ability to share that distinctiveness with each other. That's the only way I've found to feel less alone. That's the deepest way to look at the stars together. To recognize how fundamentally alone each of us is, locked in a separate body and a separate mind, and in that recognition to have the chance to feel all that reaches across that space between us, all the earth-deep connections among us that are real.

"Ready or not," the father called, taking his hand from his eyes. "Here I come!"

ACKNOWLEDGMENTS

I WOULD LIKE TO THANK Deborah Schneider for understanding the importance of this project to me from our very first conversation about it (I'd called to discuss two ideas, and after I explained this one, she didn't need to hear the second); Helene Atwan for her insightful questions and astute editing, both of which made this a better book; the production and publicity staff at Beacon for their unflagging attention to detail; Peter Canellos, Katie Kingsbury, Dante Ramos, Alan Burdick, and Peter Catapano for their guidance with the newspaper and magazine pieces that led to this book; Tanya Larkin, Cornelius Howland, and Andrew Rueb for being my trusted first readers; Tom Regan, Art Lurigio, Joyce Wexler, and David Chinitz for making me feel at home at Loyola; Nami Mun, Gus Rose, Rebecca Makkai, and Vojislav Pejovic for making me feel at home in Chicago; the Office of Research Services at Loyola for their generous research grants; Gabriela Valencia for her thoughtful assistance with research; and for all the conversations that pushed me to think more deeply about inner climate change: Ray Hearey, Alicia Pritt, James Parker, Adrienne Raphel, Rachel Kadish, Helena de Bres, Ian Cornelius, Julia Cohen, Mary Marbourg, Christopher Davis, Charlotte Porter, and Maria Robertson-Justiniano.

And a special thank you to my students at Loyola for their kindness, candor, and willingness during our workshop break to answer the question: So what are you looking at on your phone?

And thanks, as always, to my family.

NOTES

INTRODUCTION: INNER CLIMATE CHANGE

4 *Neural Darwinism:* Edelman first published the idea in 1978 in *The Mindful Brain: Cortical Organization and the Group-Selective Theory of Higher Brain Function* (with Vernon B. Mountcastle; Cambridge, MA: MIT).

5 *Edelman conceived*: From Sacks, *On the Move: A Life* (New York: Knopf, 2015).

5 *Visual cortex:* From "Rapid and Reversible Recruitment of Early Visual Cortex for Touch," by Lotfi B. Merabet et al., published in *PLoS ONE* in 2008.

6 *Tawny owl:* From "Climate Change Drives Microevolution in a Wild Bird," by Patrik Karell et al., published in *Nature Communications* in 2011.

10 *Basic structure of the human brain:* In Maryanne Wolf's 2007 book *Proust and The Squid: The Story and Science of the Reading Brain* (New York: HarperCollins).

10 *London cabbies:* From "Navigation-Related Structural Change in the Hippocampi of Taxi Drivers," by Eleanor A. Maguire et al., published in *Proceedings of the National Academy of Science* in 2000.

11 *Traditional epic poems:* William Dalrymple, in his 2009 *Nine Lives: In Search of the Sacred in Modern India* (New York: Vintage), discusses the illiterate *bhopas*, then offers this comparison: "Just as the blind can develop a heightened sense of hearing, smell and touch to compensate for their loss of vision, so it seems that the illiterate have a capacity to remember in a way that the literate simply do not."

12 *Empathy in college students:* From "Changes in Dispositional Empathy in American College Students Over Time: A Meta-Analysis," by Sara Konrath et al., published in *Personality and Social Psychology Review* in 2011.

12 *Online incivility leads to polarization:* From "The 'Nasty Effect': Online Incivility and Risk Perceptions of Emerging Technologies," by Ashley A. Anderson et al., published in *Journal of Computer-Mediated Communication* in 2014. (I talk more about the implications of the study in my 2018 piece for *Politico* "We're All Russian Bots Now.")

12 *Attention spans:* In this 2015 article from *Time*: Kevin McSpadden, "You Now Have a Shorter Attention Span Than a Goldfish," which

uses an admittedly questionable study done by the Consumer Insights team at Microsoft Canada.

12 *Receive electroshocks:* From "Just Think: The Challenges of the Disengaged Mind," by Timothy D. Wilson et al., published in *Science* in 2014.

12 *Suicide among American girls:* The CDC reported that between 2007 and 2015 suicide rates doubled among teen girls.

12 *Accidents are tied to texting:* Data reported by the National Safety Council in the 2014 edition of *Injury Facts.*

13 *neuroscientists can monitor:* Oliver Sacks, *River of Consciousness* (New York: Knopf, 2017).

13 *2,617 times a day:* From a study done by *dscout* and published in Michael Winnick, "Putting a Finger on Our Phone Obsession," in 2016.

CHAPTER 1: A MAP TO OUR MAPS

16 *Endangered Traits:* The structure of each chapter—Endangered Traits, Background, Why It Matters, Threats, What You Can Do—is a play on the World Wildlife Foundation's structure for its endangered species list.

20 *A new GPS:* This passage is adapted from an opinion piece I wrote for the *Boston Globe*, "How the GPS Defeated the Great Hippocampus," in 2014.

26 *Boone:* Chester Harding came to paint a portrait of the eighty-five-year-old Boone and claimed Boone said this to him. Versions of the statement show up in various places.

26 *One of the friends:* The description of how I came to know Oliver is adapted from an opinion piece I wrote for the *New York Times*, "Seeing Outside the Disability Box," in 2017.

28 *"Over the last few days":* The passage of Oliver's comes from his essay "My Own Life," published in the *New York Times* in February 2015, which was six months before he died.

29 *Veronique Bohbot:* Presented her findings about the use of GPS at the Society for Neuroscience's annual meeting in 2010. A good article on the effect of GPS on navigation, which discusses the work of Bohbot, Huth, and Tolman, is David Kushner, "Is Your GPS Scrambling Your Brain?," *Outside*, November 15, 2016.

CHAPTER 2: CLOCKSETTERS

36 *A man named Frederick Winslow Taylor:* Nicholas Carr, "Is Google Making Us Stupid?," *Atlantic*, July/August 2008.

39 *Living by clock time:* From Robert Levine's book *A Geography of Time: The Temporal Misadventures of a Social Psychologist, or How Every Culture Keeps Time Just a Little Bit Differently* (New York: Basic), published in 1997.

47 *Early seed germination:* From "Climate Warming Could Shift the Timing of Seed Germination in Alpine Plants" by Andrea Mondoni et al., published in the *Annals of Botany* in 2012.

48 *Sylvie Droit-Volet study:* Sylvie Droit-Volet et al., "Time Changes with the Embodiment of Another's Body Posture," *PLoS ONE* (May 2011).

48 *Alan Burdick notes: Why Time Flies: A Mostly Scientific Investigation* (New York: Simon & Schuster, 2017).

52 *David Foster Wallace:* The footnote for Wallace's article where he includes this conversation with Federer is in *The David Foster Wallace Reader* (New York: Little, Brown, 2014).

CHAPTER 3: FRAMES

56 *Tony Fadell:* "The First Secret of Design Is . . . Noticing," TED Talk, June 2015.

58 *What Thoreau would say:* Adapted from my epistolary essay "Henry David Thoreau Responds to Apple's Requests for New Apps," originally published by *Beacon Broadside* in 2017.

59 *Fischli and Weiss exhibition:* From Sanford Schwartz's review "Art That Reclaims the Ignored," published in the *New York Review of Books*, May 2016.

62 *Man in motion:* The *St. Elmos's Fire* theme song, more popularly known as "Man in Motion." Written by David Foster and John Parr, 1985.

63 *"The Summer Day":* This poem appeared in Oliver's 1990 collection *House of Light* (Boston: Beacon, 1990).

64 *Nobody sees a flower really:* Originally quoted in the catalogue for O'Keeffe's 1939 *Oils and Pastels* exhibition at Stieglitz's gallery, An American Place.

66 *Picasso said:* Dore Ashton, *Picasso on Art: A Selection of Views* (New York: Penguin, 1977).

70 *T. J. Clark:* From his book *The Sight of Death: An Experiment in Art Writing* (New Haven: Yale: 2006).

76 *More than 60 percent of Americans:* Jeffrey Gottfried and Elisa Shearer, "New Use Across Social Media Platforms," Pew Research Center, September 7, 2017.

76 *$50 million grant from Facebook:* From Jacob Weisberg's article "They've Got You, Wherever You Are," published in the *NYRB*, October 2016.

77 *Trust only our own clans:* Adapted from "We're All Russian Bots Now."

77 *A reliable way:* From Daniel Kahneman, *Thinking, Fast and Slow* (New York: Farrar, Straus and Giroux, 2011).

77 *Arendt:* In *Between Past and Future: Eight Exercises in Political Thought* (New York: Penguin, 1977).

78 *Waldron:* From his "Theoretical Foundations of Liberalism," published in the *Philosophical Quarterly* in April 1987.

81 *I wake up in cold sweats:* Tony Fadell "Nest Founder, 'I Wake Up in Cold Sweats Thinking, "What Did We Bring to the World?,"'" *Fast Company*, July 7, 2017.

CHAPTER 4: MAY I HAVE YOUR ATTENTION, PLEASE

87 *William James:* From *The Principles of Psychology*, chapter 11 ("Attention").

87 *External and internal attention:* From "A Taxonomy of External and Internal Attention," by Marvin M. Chun et al., published in *Annual Review of Psychology* in 2011.

88 *Japanese tourist drives into the Pacific:* Reported widely in March 2012.

88 *Driver in Belgium:* Reported widely in January 2013.

96 *What keeps people glued to YouTube?:* From "YouTube, the Great Radicalizer," published in the *New York Times* in March 2018.

98 *A 2014 study:* Matthew Hutson, "People Prefer Electric Shocks to Being Alone with Their Thoughts," *Atlantic*, July 3, 2014.

100 *"Just like the food industry":* Tristan Harris, "Tech Companies Design Your Life, Here's Why You Should Care," TristanHarris.com, March 7, 2016.

100 *Casinos make 80 percent of their revenue from slot machines:* Andrew Thompson, "Engineers of Addiction: Slot Machines Perfected Addictive Gaming. Now, Tech Wants Their Tricks," *Verge*, 2015.

CHAPTER 5: IDENTITY THEFT

112 *Goffman:* From his book *The Presentation of Self in Everyday Life*, first published in Scotland in 1956 (University of Edinburgh).

112 *Bourdieu:* From his book *Distinction: A Social Critique of the Judgement of Taste*, first published in Paris in 1979 (Les Editions de Minuit).

115 *James:* From *The Principles of Psychology*, chapter 10 ("The Consciousness of Self"').

122 *The Breakfast Club:* Directed by John Hughes, released in 1985.

CHAPTER 6: CONVERSATIONAL BOXES

131 *Today the average American "consumes":* Nick Bilton, "Part of the Daily American Diet, 34 Gigabytes of Data," *New York Times*, December 9, 2009.

132 *Space . . . virtually emerges out of itself:* Seamus Heaney quotes Osip Mandelstam in his essay "The Government of the Tongue" in his eponymously named book, published in 1988 (London: Faber and Faber).

134 *"Historically, 9% of Tweets":* Aliza Rosen, "Tweeting Made Easier," Twitter blog, November 7, 2017.

139 *Seventy-two to be exact:* From "Changes in Dispositional Empathy in American College Students Over Time: A Meta-Analysis," by Sara Konrath et al., published in *Personality and Social Psychology Review* in 2011.

140 *Turkle:* From *Reclaiming Conversation: The Power of Talk in a Digital Age*, published in 2015 (New York: Penguin).

144 *A serious and even combative conversation:* Remnick was quoted in the *New York Times* article "Steve Bannon Headlines New Yorker Festival," published September 3, 2018, about his intentions for the interview. He pulled the invitation to Bannon that evening.

144 *"Let them call you racist":* Morgan Winsor, "Steve Bannon: 'Let Them Call You Racist . . . Wear It as a Badge of Honor," ABC News, March 10, 2018.

144 *Apatow and Oswalt tweets:* From September 3, 2018.

145 *Havel:* From his speech "On Evasive Thinking," delivered in Prague on June 9, 1965, and collected in 1992 in *Open Letters: Selected Writings 1965–1990* (New York: Vintage).

147 *"So the best-case scenario"*: In Saunders, *The Braindead Megaphone: Essays* (New York: Riverhead, 2007).

147 *"When we go on Facebook":* Nicholas Carr, "How Social Media Is Shaping the 2016 Race," *Rough Type* blog, September 28, 2015.

CONCLUSION: LET'S MAKE A DEAL

157 *"If you were the star":* From *The Philosophy of Andy Warhol: From A to B and Back Again*, published in 1975 (New York: Harcourt Brace Jovanovich).

158 *"This is television's only trading floor":* YouTube.

ABOUT THE AUTHOR

HOWARD AXELROD'S WRITING has appeared in the *New York Times Magazine*, *Salon*, the *Virginia Quarterly Review*, *O Magazine*, *Politico*, and the *Boston Globe*. He took his undergraduate degree from Harvard University and a master of fine arts in poetry from the University of Arizona. He has taught at Harvard and at the University of Arizona, and is currently the director of the Creative Writing Program at Loyola University in Chicago. His first book, *The Point of Vanishing: A Memoir of Two Years in Solitude*, was named one of the best books of 2015 by *Slate*, the *Chicago Tribune*, and *Entropy Magazine*, and one of the best memoirs of 2015 by *Library Journal*. Author Bella Pollen said of the book: "Out of sudden and profound loss, Axelrod has drawn a haunting, tender memoir." *The Stars in Our Pockets* is his second book.